VGM Opportunities Se

OPPORTUNITIES IN
CABLE TELEVISION
CAREERS

Jan Bone

VGM Career Horizons
a division of *NTC Publishing Group*
Lincolnwood, Illinois USA

Cover Photo Credits

Front cover: upper left and lower left, Cable News Network; upper right, NTC print; lower right, National Cable Television Association.

Back cover: upper left, Jeannie Bone; upper right, Hughes Communications; lower left, Cable News Network.

Library of Congress Cataloging-in-Publication Data

Bone, Jan.
   Opportunities in cable television careers / Jan Bone.
      p.   cm.—(VGM opportunities series)
   Includes bibliographical references.
   ISBN 0-8442-4026-5 (hardbound)—ISBN 0-8442-4027-3 (softbound)
     1. Cable television—Vocational guidance—United States.
I. Title. II. Series.
HE8700.72.U6B66 1992
384.55®5®02373—dc20   92-18317
                                                                                     CIP

Published by VGM Career Horizons, a division of NTC Publishing Group.
© 1993 by NTC Publishing Group, 4255 West Touhy Avenue,
Lincolnwood (Chicago), Illinois 60646-1975 U.S.A.
All rights reserved. No part of this book may be reproduced, stored
in a retrieval system, or transmitted in any form or by any means,
electronic, mechanical, photocopying, recording or otherwise, without
the prior permission of NTC Publishing Group.
Manufactured in the United States of America.

2 3 4 5 6 7 8 9 0 VP 9 8 7 6 5 4 3 2 1

ABOUT THE AUTHOR

A prolific freelance writer, Jan Bone has been in print continuously since her first summer job in 1947 on the Williamsport, Pennsylvania, *Daily Sun*. She graduated from Cornell University, worked for trade publications in the restaurant and motel fields for several years, and began to freelance from her home in 1955. Fifteen years later, after winning a Chicago Headline Club scholarship, she returned to college, attending Northwestern University, Northeastern Illinois University, and Harper College. In 1987, she received her M.B.A. from Roosevelt University. She has been teaching writing to adults for over twenty years.

Jan's interest in cable started when her Chicago suburb asked for proposals from cable companies as the first step in the franchise process. As an elected member of the Board of Trustees of William Rainey Harper College in Palatine, Illinois (1977–1985), she realized the potential role that cable could play in expanding educational opportunities. She has been interviewed several times on the cable company's local programming—as a college board member, as a writer, and as a speaker in the public library's series on writing techniques.

Her current freelancing projects include regular magazine, minimagazine, and newsletter articles for the National Safety Council, technical writing for *P/O/P Times* (a publication covering the point-of-purchase industry), and special advertising sections for the *Chicago Tribune*. She has been the corporate profile writer for the Golden

Corridor Economic Development Commission's American Enterprise book, and has written two chapters in the National Safety Council's *Accident Prevention Manual for Industrial Operations.*

For National Textbook Company, Jan has co-authored (with Ron Johnson) *Understanding the Film: A Beginning Guide to Film Appreciation* 4th ed., 1990, a high school and college textbook which is now used internationally. In addition, she has written several books for the VGM Opportunities series on careers in film, telecommunications, robotics, computer-aided design and computer-aided manufacturing, laser technology, and plastics.

She has won the Chicago Working Newsman's Scholarship, the Illinois Education Association School Bell Award for Best Comprehensive Coverage of Education by dailies under 250,000 circulation, and an American Political Science Award for Distinguished Reporting of Public Affairs. Since 1983, she has been listed in *Who's Who of American Women.*

Jan is an associate member of the Society of Manufacturing Engineers, and a board member and treasurer of IWOCorp, the marketing arm of the Independent Writers of Chicago, the professional freelancers' association.

She is married, mother of four married sons, and grandmother of Emily and Jennifer.

ACKNOWLEDGMENTS

The following individuals were especially helpful in the development of this book: Olufunke Adebonojo, George Bodenheimer, Judy Bouer, Ruth Brumfield, Rhonda Christenson, Pat Conner, Eva Dahm, Susan Detwiler, Larry J. Fischer, Dana Fujioka, Rick Ginter, Moya Gollaher, Ralph A. Haimowitz, Barrett J. Harrison, Marc Herz, Susan Hook, Tom Hotaling, Charles Kirtley, Kathy Lane, Jeff Lee, Matt McGee, Karol Petit, Philip A. Roter, Bill Sadler, John Shemancik, Curtis Symonds, Rob Tobias, Shelly Weide, Pam Williams, and Theresa Wright.

The author also wishes to thank the following organizations for their help with this book: Baker, Scott & Co. Executive Search, Black Entertainment Television (BET), Cable Cartel, Cable Media Corporation, Cable TV Montgomery, Cabletelevision Advertising Bureau, Continental Cablevision of Northern Cook County, Cox Cable, The Disney Channel, ESPN, Excalibur Cable, Media Business Division, Heller Financial, Inc., Media General Cable, National Academy of Television Programming, National Association of Minorities in Cable (NAMIC), National Cable Television Association, National Distance Learning Center, Time Warner CityCable Advertising, Turner Broadcasting System, Inc., USA Network, The Walter Kaitz Foundation, Warner Cable Houston, The Weather Channel, WESTSTAR Cable Company, and Women In Cable.

CONTENTS

About the Author iii

Acknowledgments v

1. About Cable Television 1

The beginnings of cable television. Cable distribution by satellite. Cable's first major challenge. Cable expands. Narrowcast—audience targeting. Advertisers recognize cable's advantages. Growth pains and rapid change. Covering the field.

2. How a Cable System Works 11

Elements of a cable system. How cable signals reach a subscriber. Cable's major components. Cable systems. Program suppliers. Jobs in cable—today and tomorrow.

3. Personal Qualities 21

Flexibility. Good work ethics. Teamwork. Good self-presentation. Communications skills. Determination. Stability. Attitude.

Contents vii

4. Franchising, Construction, Financing 26
Franchising and regulation. Construction jobs. Cable finance jobs. Broadcast Cable Financial Management Association.

5. Technical Careers 37
A suburban system. A rural system. A region. Society of Cable Television Engineers (SCTE). National Cable Training Institute (NCTI).

6. Sales and Marketing 49
Marketing jobs. Marketing at two cable systems. Marketing a network. Trade association.

7. Advertising 61
Interconnects. Cox Cable San Diego. Cable Cartel. Cable Media Corporation. Technology. Reaching viewers.

8. Programming 72
Step one. A small cable system. Creating programs. Network programming.

9. Schools and Training 80
What should you study? How to find out what's available. Cutting the cost. Schools. Seminars. Internships.

10. Job-hunting Tips 95
"People skills" are vital. Using trade publications effectively. Do your homework. Presenting yourself at your best. If you're a cable veteran. What to expect.

11. New Technology 105
Fiber-optic cable. Interconnects. Technologies merge. Analog and digital technology. High definition television (HDTV).

viii   *Opportunities in Cable Television Careers*

12. Women and Minorities **110**

Minority employment. Increasing role for women. A station general manager. A regional director of a pay service. Women In Cable. A cable system manager. Walter Kaitz Foundation. A network vice-president. The National Association of Minorities in Cable (NAMIC).

13. International Markets **129**

Cable television in Canada. Cable television in Europe.

Appendix A: Associations **136**

Appendix B: Recommended Reading and Resources **139**

Appendix C: Glossary **145**

CHAPTER 1

ABOUT CABLE TELEVISION

Probably we shall never know exactly how the first cable system originated shortly after World War II. There's a story in the industry that a gas station owner, tired of poor television reception in the Pennsylvania mountains, climbed the hill behind his station, put up an antenna, and ran a wire down the mountainside to his television set. There's another story about the owner of a local radio station in Oregon building the first cable television system in 1949.

However cable television started, it has proved popular with viewers, and its popularity is increasing rapidly. In 1990 an opinion poll was conducted by the Roper Association for the National Cable Television Association (NCTA), the major trade association which represents the nation's cable television industry. The poll found that the public viewed cable's programming as "superior, in quality and variety, to that offered by traditional broadcast television, whether cultural, children's, sports, or general entertainment programming."

By 1991 *The Broadcasting Yearbook* reported there were 10,823 operating cable systems in the United States, serving 28,798 communities. Another 200 franchises had been approved, but not built. Although some systems had fewer than 100 subscribers, the largest, Cablevision Systems in Oyster Bay, N.Y., had more than half a million subscribers.

Just over 60 percent of U.S. households subscribed to cable television in 1991, according to Nielsen Media Research—a 27.2 percent increase since 1985, and a 3 percent gain over 1990, when Nielsen figures put the cable television subscriber count at 56.1 million U.S. homes. That figure roughly translates to more than an estimated 158 million viewers.

However, cable penetration—what percent of households have cable—varies among markets. Nielsen Media Research for May 1991 credits cable with 70.9 percent penetration in Boston, 60.2 percent in Chicago, 56.6 percent in Los Angeles, and 47.1 percent in Dallas-Fort Worth.

THE BEGINNINGS OF CABLE TELEVISION

It wasn't always so, however. At the beginning, most people thought of cable as just a way to get better television reception, especially if they lived considerable distances from network broadcasting stations. When mountains or forests got in the way of television signals, cable became a means for getting better reception. Most cable operators merely put up a central receiving antenna to pick up programs from broadcasting stations. In fact, cable was called Community Antenna Television (CATV). Later, cable operators began to use microwave radio relays for signals. As television became less of a novelty and more an accepted part of America's way of life, cable's popularity grew.

By 1960, says Cobbett Steinberg, author of *TV Facts,* there were 650,000 subscribers and 640 operating cable systems. The new-fangled idea had worked. But not everyone was happy.

One subsidiary of Paramount Pictures tried a pay-TV experiment in Etobicoke, a Toronto suburb. Subscribers dropped coins in a meter on the set; movies cost $1, and for $1.50, they could watch the Toronto Maple Leafs' away-from-home hockey games. After four years, however, the system shut down—costing its owners more than $6 million in losses.

California tried its own pay-TV experiment in the early 1960s. The plan, called Subscription TV, Inc., carried baseball games of the Brooklyn Dodgers and the New York Giants, newly settled in their respective homes of Los Angeles and San Francisco. The news so infuriated the California public who'd been looking forward to seeing "their" teams on network TV that in 1964, a statewide referendum actually outlawed pay television. Later, the state supreme court decided that the referendum was unconstitutional. By that time, however, the whole idea had been shelved.

CABLE DISTRIBUTION BY SATELLITE

To a certain extent, cable was still a useful toy. Subscribers liked the better reception it offered, but broadcast networks weren't taking the competition seriously—then.

Now they do, as cable loyalty grows.

In 1979–80, according to the Cabletelevision Advertising Bureau* (CAB), during primetime, 87 percent of all U.S. households with television watched broadcast networks. By 1990–91, prime-time broadcast network share had dropped to 60 percent. Daytime television viewing showed a similar trend. In 1979–80, 75 percent of all U.S. households with television watched broadcast network stations. By 1990–91, only 51 percent were doing so.

A CAB analysis of Nielsen Home Video Index data from October 1990–June 1991 showed that advertising-supported cable TV networks captured a 31 percent combined share of the prime-time television audience in the 56 million U.S. homes that receive cable. For those months, during prime time, basic cable networks gained 1.08 million households per average minute from a year earlier, compared with an increase of 58,000 households for independent TV stations and a loss of 394,000 households among the major broadcast network-affiliated stations.

What percentage of viewers watch broadcast networks vs. cable television?

The answer depends on how you define "viewers."

---

* Cabletelevision Advertising Bureau is a national trade association for making cable more valuable as a national, regional, and local advertising medium.

For *all* U.S. households that have television (including those who don't have cable), on an average 24-hour day in 1990–91, 36 percent watched broadcast network dayparts; 30 percent watched cable; 21 percent watched independents; and 17 percent watched non-network dayparts.

But if you look at only the households which had cable television, on an average day in 1990–91, 44 percent watched cable; 31 percent watched broadcast network dayparts; 17 percent watched independents; and 15 percent watched non-network affiliates.

In one quarter-hour on September 5, 1991, USA Network's coverage of the U.S. Open, a tennis tournament, exceeded audiences for both ABC and CBS in cable homes in that time period.

Cable's role in today's life-style has become increasingly important. For instance, CAB's 1991 figures showed that Americans spent more for cable ($18.2 billion) than on movie attendance ($4.5 billion) and videocassettes ($11.0 billion) combined.

CABLE'S FIRST MAJOR CHALLENGE

The first major challenge to broadcast networks came in September, 1975, when HBO* went "up on the bird," as one cable expert described it. Time, Inc. decided to commit over $6 million to lease two channels on RCA's domestic satellite, and to distribute pay cable by satellite, starting in September, 1975.

These orbiting space stations can relay signals from one point on earth to another. HBO's use of the satellite (it had the "bird" to itself for over two years) positioned it as an industry leader. Soon, however, other networks began forming. They would supply programming to cable systems, using satellite technology.

Today, distance is no obstacle for television or cable. Millions of Americans watch the French Open or Wimbledon tennis tournaments; still more watch the Olympics. Television even brought the 1992 Super-

---

*HBO (Home Box Office), a pay cable channel featuring movies, concerts, and specials.

Bowl—admittedly, not till five weeks after the actual event—to viewers in the former USSR, complete with commentary by two ex-Soviets who struggled to explain such terms as "tight end" in Russian.

Through its immediacy and flexibility, cable television changes our perspective on world affairs by bringing us news *as it happens.* Cable is doing this—not only for the "average" viewer, but also for world leaders.

Time reported that during Desert Storm, the short-lived war launched in 1991 to protest Iraq's invasion of Kuwait, on the night that the bombs began to fall on Baghdad, virtually every senior official in virtually every government was watching CNN. As *Time* put it, "From Rome to Riyadh, London to Lagos, Beijing to Buenos Aires, Cable News Network is on more or less continuously in the suites of a vast array of chiefs of state and foreign ministers. What computer messages can accomplish within an office, CNN achieves around the clock, around the globe."

CABLE EXPANDS

By the mid-1970s, cable was beginning to expand. Roughly 15 percent of America was wired, and satellite services began to develop. To receive the signal from a communications satellite, the cable TV operator has to have an earth station—the familiar satellite receiving dish. Once the operator gets the program, the signal is retransmitted across coaxial lines directly to each cable subscriber.

As one of the first networks to use the satellite, ESPN began telecasting full-time on September 7, 1979, concentrating on sports events and sports news. Gradually, it increased its programming schedule until, a year after its first broadcast, it offered 24-hour-a-day service.

On December 17, 1976, cable's first SuperStation was launched, when programs shown on WCTG, owned by Turner Communications Group, were beamed via satellite to cable homes nationwide. Less than four years later, Turner launched Cable News Network. Initially started with 1.7 million subscribers, the world's first 24-hour, all-news network now reaches 93 countries on all continents. In the United States,

60 percent of America's 92 million television households receive CNN, and the number continues to grow.

NARROWCAST—AUDIENCE TARGETING

The availability of satellite services made a big difference in the way cable developed. Suddenly there were more stations to watch, and more programs to see. Cable began to *narrowcast,* a technical term the industry uses to describe programming aimed at special, fragmented target audiences. (Broadcasting, in its literal sense, means a single program aimed at the wide audience.)

As an example of narrowcasting, VH-1, the "other" video channel from MTV, changed its format in 1991, positioning itself to provide the greatest video hits for a baby boomer target audience—viewers in the coveted 25–49 age group. As a result of its new makeover format, VH-1 recorded a 25 percent increase in ad sales, reaching more adults 18–49 with household incomes of over $40,000 than any other basic cable network, MTV executives said.

The narrowcasting approach extends not only to cable itself, but to the methods marketers for multiple-system operations (MSO) use to attract subscribers. Instead of sending out the same direct mail piece emphasizing ESPN's National Football League schedule to nonsubscribers, United Artists Cable's Jerry Maglio says, "I'd like to produce a special piece highlighting the Consumer News and Business Channel to the subscription lists of *The Wall Street Journal, Forbes,* and *Fortune.*"

Market researchers track cable audiences, keeping score on just who's watching what. For instance, in 1992 the basic cable kids' network, Nickelodeon, announced that "Clarissa Explains It All," a live-action sitcom about a 14-year-old blonde tomboy in jeans, was tremendously popular with young males. About 70 percent of the show's viewers ages 2–11 were male, Nickelodeon said. In fact, among Nickelodeon households, the show had a higher concentration of males ages 6–17 than either "Beverly Hills 90210" or "Fresh Prince of Bel-Air," two youth-oriented shows on other networks.

Viewer numbers translate into dollars when combined with cable's narrowcasted, segmented audience.

According to the *A. C. Nielsen Cable Demographic Report* (second quarter, 1990), Headline News, the round-the-clock program that delivers the latest news in 30-minute segments, attracted more adults 18–49 with incomes over $60,000 than any other basic cable network in that time period. Headline News, which analyzes audience composition, says its viewers are 39 percent more likely to hold a professional or managerial job, and 38 percent more likely to have completed college than the U.S. average.

Those demographics translate into spending patterns. In 1990, Headline News viewers ranked far above the national average in having health club memberships, owning two or more cars, using credit cards, owning VCRs, and traveling. Statistics like these gave the network ammunition for pitching its services to potential advertisers.

ADVERTISERS RECOGNIZE CABLE'S ADVANTAGES

Since the early days, advertisers have taken note of cable—the new kid on the block. Instead of paying for television broadcasting with the three big networks alone, they found that they could present commercials aimed at specific audiences watching special-interest networks. In 1980, ad revenues for cable were approximately $45 million; by 1982, ad revenues for cable reached $241.8 million, according to industry expert Paul Kagan. Just 10 years later, according to the Cable Television Advertising Bureau, ad revenues for cable were $3.55 billion.

However, cable still ranks substantially below broadcast networks in total numbers of dollars spent on advertising. Despite cable's $3.5 billion ad revenue, 1991 broadcast revenues, including network, spot, and local TV, hit the $16.4 billion mark—$10.16 billion of that coming from advertising on CBS, ABC, and NBC. Cable, however, had an 18 percent rise in ad revenue over the preceding year, compared to a 4.5 percent increase for broadcast networks.

Because the cable industry is growing and changing, no one is sure just how much of the advertising dollar will eventually go to cable and how much to broadcast networks. Paul Kagan Associates, Inc. projected revenues for cable television advertising would reach $5.2 billion by 1995. Analysts such as Kagan track these figures carefully; their results are studied by ad agencies and businesses before advertising moneys are budgeted.

Cable network advertisers include top national companies. The Cabletelevision Advertising Bureau says that 1990–91 expenditures for six measured networks (CNN, ESPN, FAMILY, MTV, TBS, and USA) included $62.7 million by Procter & Gamble, $35.4 million by Time Warner; $31.0 million by General Motors; $29.2 by Anheuser-Busch; and $19.2 million by AT&T. Activity on additional cable networks, CAB said, could have increased these expenditures by as much as 50 percent.

GROWTH PAINS AND RAPID CHANGE

As a relatively young industry, cable is volatile. The network that's here today may be gone by tomorrow. Or the network that doesn't exist today may have millions of viewers, seemingly overnight.

Even smooth-running networks have complicated histories. Take, for example, USA Cable Network, a 24-hour-per-day, advertiser-supported service seen in more than 58 million households as of January, 1992. The network began in 1977 as the Madison Square Garden Sports Network, originally carrying only events taking place in the Garden. Shortly thereafter, it expanded to include sports events from coast-to-coast and overseas.

By April, 1980, MSG Sports and United Artists-Columbia became partners, forming USA Cable Network. Eighteen months later, the network was bought by Time, Inc., Gulf & Western's Paramount Pictures Corp., and MCA, Inc. In 1987, Paramount and MCA became equal partners in their ownership of USA Network.

A year later, USA network announced it had commissioned a single-season schedule of 24 original, exclusive World Premiere movies—the largest original production commitment to that point in the history of cable. In 1988, the network also set aside a budget of $50 million for marketing of new programming, and opened a new in-house post-production facility in mid-Manhattan.

USA network continued its expansion plans. By September 1990, the network had agreed to become exclusive ad sales representative for Super Channel, the largest European cable network with 23 million subscribers in 21 nations. And in August, 1991, the network and *McCall's Magazine* announced a joint cross-media advertising opportunity using select USA World Premiere movies—an opportunity for advertisers that allowed them to buy both broadcast time and print advertising in a combination package.

COVERING THE FIELD

Because the cable scene is changing so rapidly, you almost need a scorecard to know the players. That's why, when you're considering a cable career, it is *essential* you keep up with what's happening and who's where. Magazines like *Cable World* run monthly columns listing key people and their job changes. *The Hollywood Reporter* regularly runs production information that includes names and addresses of networks and independent production companies that make films for cable. Other publications provide subscribers with franchising, construction, and licensing information throughout the country.

Media analyst Paul Kagan covers the cable industry intensively. His monthly newsletters include *Cable TV Advertising, Cable TV Banker/Broker, Cable TV Investor, Cable TV Investor Charts, Cable TV Law Reporter, Cable TV Programming,* and *Cable TV Technology.*

Confusing? Probably. It's hard to sort it all out, and to know just what developments like these can mean if you're considering cable as a field for *you*. In fact, one cable expert puts it bluntly: "There are no careers in cable," she warns, "just jobs in cable." Still, she does not mean to be discouraging. "Cable needs good, capable people," she feels, "people who can understand its potential and the exciting future it holds."

The National Cable Television Association (the chief trade association in the industry) says approximately 91,000 persons were employed in cable in 1991, up from the 25,000 who had cable jobs in 1975. And they say employment is growing.

Do *you* want a job in cable? Can you get one?

These are the questions that *Opportunities in Cable Television* wants to tackle. There are no easy answers in today's tight job market. But this book will give you information about cable's various components, places to go for more details, and (through personal vignettes) an awareness of how a few people in the industry view their jobs.

To most of them, being on the job is not really work. Sure, there's a job to be done. It's pressured with deadlines. Television can't wait; the story must be aired as soon as possible. Television is voracious, chewing up scripts, programs, and even old movies with a monstrous appetite. But cable is something special—it has its own special character, values, and opportunities for those who work in the industry and love it. Maybe it will be special for you, too.

CHAPTER 2

HOW A CABLE SYSTEM WORKS

Cable is a wire that carries video and audio signals directly into the homes of subscribers. Older cable systems had limited channel capacity. For instance, in 1978, only 25 percent of cable systems had the capacity to carry more than 12 channels. But today's technology makes it possible for *new-build* systems to be constructed with over 100-channel capacity and interactive capability.

Factors that determine the size of the cable facilities include the number of people in a community and how sophisticated the community wants its cable system to be. Many cable systems installed 15 to 20 years ago are upgrading their facilities with newer technology. *Rebuilds* are often done to expand channel capacity, satisfy franchise requirements, and replace gradually decaying coaxial cable with fiber-optic cable.

The switch frequently improves service to subscribers by making signal quality better. In fact, cities considering renewing cable franchises frequently demand that cable systems improve broadcast signal quality. However, at this writing the Federal Communications Commission is in the process of setting rules to establish industry-wide technical standards.

At Cox Cable Roanoke (Virginia), the 1992–94 rebuild plan replaced 550 miles of cable and upgraded the remaining 410 miles. The fiber-to-the-service-area rebuild increased cable capacity from 36 to 62 channels. In addition, an institutional network was created to link 21 municipal and educational sites.

Cox Cable Communications planned the project carefully. It wanted to keep options open for future business expansion, but to do so without locking into any fixed technology.

ELEMENTS OF A CABLE SYSTEM

Principal elements of a cable system are a *community antenna, satellite downlink,* a *local studio, signal-processing equipment, amplifiers, line extenders, drops,* and the *coaxial* or *fiber-optic cable* itself.

You'll find more about jobs building cable systems in chapter 4 of this book. Much of the United States is already wired for cable; that is, by 1992, over 60 percent of the country's television households had cable. But in areas such as new subdivisions, building a cable system can take many months. If the cable will be carried on telephone poles, each telephone or utility pole in the community must be inspected and mapped. Old poles may have to be replaced by the utility companies. Phone or utility lines may have to be moved, and new crossbars may have to be added. All this must be done before the cable company can begin to install its equipment.

If the cable system is to be installed underground, a similar procedure is followed. Existing utilities are mapped thoroughly, and plans must be made for the digging and laying of cable so as not to disrupt utility services. Cable installation is a costly business. Because of the complex maze of existing utilities in large cities, the task—and the cost—are much greater.

When cable sites are finally mapped and prepared, cable construction workers, or installers, begin laying down the coaxial or fiber-optic cable.

In addition, earth receiving stations, known as dishes, are constructed to receive signals from the communications satellites that are in orbit about 22,500 miles above the earth.

Satellite technology became commercially feasible in 1973. In fact, delegates to a cable conference being held in Anaheim, California, watched as programs were beamed from an East Coast transmitter to a communications satellite, which then sent the signals on to the first cable earth receiving station. For the premiere viewing, they saw a sports event from Madison Square Garden, political speeches from Washington, D.C., and short films.

Satellites serve cable systems throughout the country simultaneously, 24 hours per day. The communications satellites that relay the programming are placed in a geostationary orbit. They rotate about the earth at the same rate the earth revolves. They stay in the same position with respect to the cable stations they serve. There are over 25 satellites in the Western Hemisphere.

The satellites, in effect, become "antennas" in receiving signals from the program source and transmitting them to any earth stations below that can pick them up.

Each cable system constructs or installs "dishes" to receive the programming available from various satellite programming services. These can include SuperStations like WTBS (Atlanta), WOR (New York), and WGN (Chicago).

HOW CABLE SIGNALS REACH A SUBSCRIBER

Local cable systems (who already have studio and office facilities in place and who long ago made arrangements to receive program signals) don't have long to wait before they can serve their subscribers in new-build areas.

First, of course, drop lines must be placed to connect the main cable and the subscribers' television sets. Feeder lines connect the drop lines to the individual houses. Households that sign up for cable are visited by installers, who will bring the cable through the wall into the house and hook it to the television set. The cable company supplies a small box (the convertor) which is placed on top of the television set. The set itself remains tuned to a particular channel, while the subscribers pick their desired programs by changing the channel number on the covertor.

With addressable convertors (a sophisticated version of the box on top of a subscriber's set), changes in programming can be accomplished easily. If the system is interactive, a subscriber can order a pay-per-view event, such as a movie or a championship prize fight, right through the convertor. The converter "knows" who watched, and bills subscribers accordingly.

Earlier in cable's history, most cable systems offered a broad line of basic services, supplemented by a handful of full-priced premium services. Today, and in the foreseeable future, the distinctions between "basic" and "pay" are becoming increasingly blurred. Limited-basic and expanded-basic tiers, and the mini-pay Flix offered by Showtime in 1992, as well as "the coming reality of a smorgasbord of à la carte offerings" (as Paul Kagan senior analyst Larry Gerbrandt puts it), have helped erase those distinctions.

CABLE'S MAJOR COMPONENTS

To those outside the industry, cable appears to be much like network television broadcasting. You turn on your television set, switch to your favorite cable channel, and start watching your program. It sounds simple and obvious. Presumably, then, jobs in cable have to do with producing that program and getting it to your television set so you can watch it.

Cable is divided into (a) companies concerned with the engineering necessary to get that program signal to you, with selling and marketing their particular packages of signals, and with the business functions necessary for handling your subscriber fees and making a profit (cable operators); (b) companies that produce the programs (program suppliers); and (c) companies that provide services associated with cable, such as legal firms that deal with franchising; market research firms that provide information on demographics; financial services companies; and consultants. Average cable subscribers don't see or hear much about those service-providing companies; nevertheless, they represent another segment of cable employment that you may be able to tap into.

It's true, as the National Cable Television Association estimates, that about 90,000 people work in cable. But whether that cable work is with *cable operators*, with *program suppliers,* or with *"outside" service providers* affects the number of jobs and job openings, and the training required.

CABLE SYSTEMS

One possible way to break into cable, many experts feel, is to start with a cable system. Operating companies need sales and marketing persons, technical experts to keep the system running and handle emergencies, advertising persons to sell commercial time, and personnel such as customer representatives and accountants, financial, and secretarial workers.

With few exceptions, cable systems operators find it's cheaper and easier to buy professional programming than to try to produce it themselves. The particular programming mix suitable for each town or city on the cable is almost always set by the local operator. Sometimes the franchise terms determine what programs can be shown; i.e., there may be a ban on X- or NC-17-rated movies, or only certain times when R-rated pictures can air. Many times, the local operator knows the community, and target-markets the type of programming mix that will sell most easily, given viewer demographics.

The mix of channels a particular cable system offers today may not be what you'll see if you tune in next month. Cable operators are sensitive to the demands of the market—that is, they want to increase the number of viewers (and therefore, the rates they can charge for local advertising). Consequently, if market research shows that a channel or group of channels isn't truly profitable, the cable system will add or delete channels according to what they think viewers want to see.

Often, cable systems package certain channels together, much as a restaurant might put together combinations of foods for featured meals.

For instance, Continental Cablevision of northern Cook County (Illinois), headquartered in Rolling Meadows, about 35 miles northwest of Chicago, serves nearly 50,000 subscribers in the suburban communities of Rolling Meadows, Buffalo Grove, Elk Grove Village, Palatine, and Hoffman Estates. In 1992, the system offered new subscribers several choices.

The standard premium package included standard cable service, a free wireless remote control, and two premium services. Subscribers could select any one of four combinations:

HBO-Cinemax
HBO-Disney
HBO-Showtime
HBO-The Movie Channel

Subscribers could pay a moderate additional amount to add a third premium service.

Continental Cablevision also offered a standard cable-ready premium package, which gave viewers standard cable-ready service, and a combination of HBO and The Movie Channel. The Limited Premium Package (another option) included limited cable service and a choice of two premium services. A fourth choice, the Limited Cable-Ready Premium Package, allowed subscribers limited cable-ready service and a combination of HBO and The Movie Channel.

PROGRAM SUPPLIERS

The major source of programs for the various cable systems is the satellite-delivered networks. An excellent source of information on those networks, *Producers' Sourcebook: A Guide to Cable TV Program Buyers,* can be purchased from the National Academy of Cable Programming, 1724 Massachusetts Avenue, NW, Washington, D.C. 20036. The 2,500-member Academy, which develops the annual national and local CableACE competitions to honor the best in cable television programming, also coordinates (with NCTA) National Cable Month.

The book, updated frequently, lists the national networks and background facts on each of them: hours programmed, number of subscribers, target age groups, and parent companies. Reading the book will give you an indication of each network's interest in buying independent programs. For instance, in 1992, Comedy Central purchased approximately 1,300 hours of original programming; The Disney Channel required independent producers to submit work through their respective agents; and Lifetime Television had a "special interest" in programming that dealt with parenting, exercise, cooking, magazine programs, and documentaries. You would have learned, also, that Lifetime Medical Television was the world's largest producer and distributor of programming for physicians and other medical professionals, and that Healthlink Television, a healthcare information system shown in doctors' offices, was a video service offering healthcare education to patients in the waiting rooms of pediatric, obstetrical/gynecological, and family physicians.

Different Types of Networks

Two different types of satellite networks exist: (1) pay, or premium, television, and (2) ad-supported satellite networks.

Pay, or premium, networks are subscriber-supported. That is, subscribers pay an additional fee to their local cable system in order to be able to watch them. Usually there's a per-channel charge. Often, local systems package two or more of the pay-channels together, as Continental Cablevision of Northern Illinois has done.

These networks include HBO, Showtime, Cinemax, The Movie Channel, and The Disney Channel, all of which offer first-run motion picture services. Some of them also have made-for-television productions, serials, adult soap operas, and nightclub acts.

Other pay-television services include Playboy At Night (adult programming) and Encore (which offers hit movies from the '60s, '70s, and '80s).

Ad-supported networks and satellite services are programs produced by the satellite networks on which advertisers sell time. Some of these commercials are bought by national advertisers, and are shown on the particular local systems that choose to carry an individual satellite service. Others, frequently handled through interconnects, may have commercials that appear on individual systems. The interconnects make it possible to sell local advertising to regional and national marketers who would otherwise not buy local system purchases.

Some ad-supported networks may be offered as part of the basic cable service on a particular local system. They include The Discovery Channel, Nickelodeon, and Nick at Nite (Nickelodeon's nighttime programming), TNN (The Nashville Network), USA Network, and Turner Network Television (TNT).

Sometimes these networks offer specialized, narrowcast services, reaching for specific target audiences in an attempt to persuade advertisers of particular products that they can deliver to an audience that's likely to buy.

The Weather Channel is an example of specialized programming. It's owned by Landmark Communications, one of the larger media companies in the United States, with interests in newspaper publishing, broadcasting, cable television systems, special interest publications, and cable television programming. The Weather Channel, which began on-air shows in May, 1982, reached 51 million homes ten years later.

Although The Weather Channel presents text and numbers displayed on the screen—material showing temperature and brief forecasts—the heart of its 24-hour programming is specialized forecasting by trained meteorologists. One minute of every five minutes of cablecast, however, is devoted to viewers' *local* weather reports and forecasts. Another minute of every five is devoted to a commercial message. Ten of those ads per hour are originated by The Weather Channel network, while two of the commercial minutes are reserved for sale or use by the local cable operator.

A technical breakthrough in satellite technology made The Weather Channel feasible, since it can totally localize the information it presents. The network supplies The Weather STAR (Satellite Transponder Addressable Receiver) to each affiliate cable system. The Weather STAR receives, stores, and displays automatically upon command from the network—a wide range of local forecasts, local weather warnings, bulletins and weather statements.

Forecasts for the local community served by each cable system are addressed to The Weather STAR at that cable system. They are transmitted in the *vertical blanking interval*. The television picture is made up of 525 scanning lines. But because certain of these lines are not used to make up the picture the viewer sees, they can be used to transmit other information.

The vertical blanking interval, then, includes the lines that make up the unused portion of the picture between the frames of active video. Out of the high-speed data stream being transmitted, The Weather

STAR at a particular cable system's *headend* "recognizes" its own address and captures the individual local forecast.

In addition, The Weather STAR at each cable system receives and instantaneously displays weather warnings, bulletins, and statements. It will interrupt all regular programming, even commercials, to display them.

The Weather STAR also inserts local commercials and clients, and performs all real-time switching functions. That is, it automatically switches from national programming to the local forecast upon command from The Weather Channel network in Atlanta.

JOBS IN CABLE—TODAY AND TOMORROW

Are there jobs in cable? Can you get one?

That depends.

On the cable operations side, jobs with any of the 10,823 systems largely center around getting and keeping more cable customers, increasing the revenue-per-customer by developing and selling combinations of premium cable services, and in getting the cable signal to the customers. Jobs also exist because of the accounting and clerical functions required to keep the system running and profitable. Cable systems need skilled, reliable technicians and supervisors who can keep the system running (and fix it quickly if there's an outage), who can upgrade their skills, and who can keep up with rapidly changing technology.

As the world of "telecommunications" broadens to include cable, former distinctions are becoming blurred. Cable is beginning to expand into the telephone business, and telephone companies (telcos, for short) are hoping to expand into the cable business. Cable systems, set up originally on analog technology, are becoming increasingly digital in the evolving environment. The spread of digital technology will almost certainly open up more opportunities for cable jobs.

On the programming side of things, the exploding growth of cable has meant a continuous need for programs. Whether a made-for-TV movie has been commissioned by a cable network, or independent producers have convinced suppliers to purchase shows they've developed, quite a few persons have been involved in getting those programs ready for viewing.

It's unrealistic to think you will jump straight from being a communications major or television production major in college to a programming job with a major network. However, independent production companies (check *The Hollywood Reporter* for names and addresses) may have openings at a very junior level; getting experience *anywhere* can help you in a cable career. In addition, most cable systems train volunteers in cable production. Those who've been checked out as proficient on equipment can frequently write, develop, shoot, and edit programs shown on cable's local access channels—gaining experience and a reel of video clips that will help demonstrate what they can do.

Under the broad heading of "cable-related services," many companies provide what cable television needs: from equipment rental and leasing, film labs, and insurance services, to the publishing of cable guides, providing of billing/accounting systems, and the translating of foreign languages.

Check *Cable Yellow Pages*—it's an annual phone directory "for and about the cable television industry." You can purchase the book from Media Image Corporation, 6060 S. Willow Drive, #312, Englewood, Colorado. The book lists products and services by vendor category. You'll have to do your own research on companies, learning if they have openings and what you need to do to qualify for an entry-level job. However, chances are you've never suspected the scope of behind-the-scenes business opportunities that back up cable television operations.

Without exception, everyone interviewed for this edition of *Opportunities in Cable Television Careers* believes that those who truly want to work in cable *will* make it—if they're flexible, willing to do what is necessary to "get in," eager to learn, and capable of changing as the industry matures and develops.

Glenn Jones, CEO of Jones Intercable, Inc., calls cable "an evolving, dynamic, youthful, and complex industry." If being part of that industry is your dream, there's a lot you can do to help make that dream come true!

CHAPTER 3

PERSONAL QUALITIES

What personal qualities are necessary to succeed in cable television? What do people with jobs in cable feel are the advantages and disadvantages of working in this expanding field?

FLEXIBILITY

"The cable industry is still young enough that the book hasn't been written on it," says Philip A. Roter, vice-president of sales and marketing for Cable TV Montgomery in the Washington, D.C. suburban area.

> Someone young can still come into cable with a good idea and have that idea rolled out in a week if it makes good business sense. No one is going to say, 'They did it this way 20 years ago, and that's why we have to do it this way today.'
>
> Cable is constantly evolving. It's fast-paced and ever-changing. Some people are more comfortable with change than others. The ability to accept change—to learn and grow with it—helps.

Susan Hook, general manager of WESTSTAR, the cable system that serves Bishop, California, agrees.

> Cable is young enough that it's still changing really fast. You have to be able to handle change. If you think you're going to get into a job where everything stays the same—you're not.

Cable is still at the whim of the politicians, both local and federal. Exciting technology is just around the corner! Being comfortable with change is an asset in cable.

GOOD WORK ETHICS

Cable systems are willing to promote from the ranks, *if* persons have proven themselves.

"Are you just going to show up for work on time, or are you also going to do a good job?" asks Charles Kirtley, director of marketing and new development for Excalibur Cable.

"When we hire, we're looking for experience and excellent references. We check them."

At the Bishop, California, system, Hook looks for stability in her job force. "We do drug and alcohol testing," she says. "We look for a drug-free background. We look for a clean driving record."

Matt McGee, 23-year-old news director at Hook's cable station, points out that "people hire you because they think the new employee can make their job easier. Don't disappoint them."

He doesn't. As Hook puts it, "Matt can take a camera out, go film an interview, edit the piece, and then deliver the story on-air. Fast."

TEAMWORK

Interpersonal skills are close to the top of the list when cable executives are asked which personal qualities they value highly. They feel the ability to work with others is essential. As one veteran cable reporter puts it, "Television is a team enterprise."

> In print journalism, you have more control of how you handle a story. But in broadcast television and cable, you depend on many people to make it work. For instance, an associate producer may help me locate photographs and documents to illustrate stories, and even do some of the on-camera interviews.
>
> Even scheduling the assignments requires teamwork because we share our cameras with the other reporters. The footage my camera

crew shoots will be sent to network headquarters and put together there for tonight's newscast.

GOOD SELF-PRESENTATION

Salespersons who call on prospective cable customers must make a good appearance, cable experts believe. They must be able to learn what the services are, and to describe them accurately to prospects. They must handle themselves well and be articulate.

"It's easier to teach someone to turn a wrench than to teach them how to talk to people," warns Marc Herz, vice-president of sales and marketing for Warner Cable Houston.

> Communication skills are important. So is knowing how to sell and how to present yourself.
>
> Oftentimes, you're the front line of the company's presence in the community. You create the company image. If you're not courteous and respectful, subscribers get the impression that the cable company doesn't care about them.
>
> We do want people to get work done in a time-efficient manner, but we're also asking them to do quality work in a short time. We want people who will know how to pay attention to the quality of their interaction with customers.

COMMUNICATIONS SKILLS

Customer service representatives (CSRs) have a demanding task. Frequently they handle over a hundred calls a day, taking the next one in sequence almost immediately after they've disconnected from the preceding customer call. Because a number of the calls will be complaints, CSRs must be able to probe for information, listen carefully, and provide instruction over the phone. Above all, they need lots of patience. "We get many calls where people don't know they're supposed to be tuned to channel 3," says one cable manager.

> They turn to channel 4 and call us because their set isn't working.

> Representatives ask, 'Is the set plugged in? Is the convertor plugged in? Is the tuner on channel 3? Has the set been moved recently?'
>
> Customers tell you, 'I'll go check. Wait a minute. The set's in the other room.' Then they come back. 'Yes, it's plugged in. Channel 3? Well, I didn't look. I'll have to go back.'

Customer service representatives must be able to do that kind of phone coaching. If they can't resolve the problem, they dispatch *demand maintenance technicians* to get the customer's set back in operation.

DETERMINATION

Determination is another quality that's necessary if you want to succeed in cable. Because many stations and networks are on the air seven days a week, 24 hours a day, you've got to be willing to work, no matter when you're needed. You may be putting in long hours, or working on weekends. You'll be expected to do whatever it takes to get the job done.

Technical people in cable frequently are on standby, and usually receive extra pay just for being available, whether or not they are called to work.

STABILITY

Stability is another quality that cable companies value highly, especially for technicians. One cable veteran feels technicians tend to move from job to job. Experience with too many cable systems is a warning flag on a resume, he feels.

Personnel managers agree. So much on-the-job training is necessary for cable employees that they don't like taking chances on "here-today, gone-tomorrow" applicants. Indeed, though they value experience, they prefer to hire persons whose backgrounds show professional growth with one or two companies, rather than a pattern of job-hopping.

Theresa Wright, employment specialist at Media General Cable of Fairfax (Chantilly, Virginia), puts it this way:

Qualities in any new hire we look for include a good track record, tenure, and stability. We demand excellent oral and written communication skills, proven creative and follow-through abilities, time management and decision-making skills, and a high level of professionalism.

ATTITUDE

"If you're coming out of college, looking for a job in cable, don't turn up your nose at the mailroom if that's what you're offered," warns ESPN's Rob Tobias. "That may be the only job available.

"Take it. Work hard. For every opportunity you're given, prove yourself and you'll be noticed. Getting that foot in the door takes perseverance and humility."

Here's how one cable veteran describes the key to success:

> If you had to pick a common denominator of the most successful people who have risen to the top of the industry, they're people that nothing can upset. Whatever goes off the track, whatever problems they encounter, they're almost imperturbable. They have the ability to keep going forward on a mission. They're visionary. They think of different ways to get the job done—ways that haven't been tried before. They are self-starters.
>
> Being a person who sits at a desk, doing only what's on the desk and in the job description, is probably the opposite of being a growing member of the cable industry. Being successful in cable, and therefore in the communications revolution that's already under way, takes adventuresome, intellectual curiosity. You have to be thinking where you and the industry will be tomorrow. There's a lot of new territory to chart!

CHAPTER 4

FRANCHISING, CONSTRUCTION, FINANCING

Virtually all cable systems are regulated by the franchising process. The first step in starting a cable system is to get a franchise. Broadly speaking, each municipality or local governing body which the cable system serves has the power to regulate it, and to set the terms by which the system can operate. In areas that are unincorporated, a county or township board usually has this power.

Requirements vary widely from community to community. They're set by the elected officials (or their representatives) under whose jurisdiction the cable system falls. Sometimes village trustees or city aldermen appoint a cable commission to do the preliminary work and make recommendations before a franchise arrangement is drawn up.

When a municipality is ready to consider cable, it invites cable companies to submit applications by placing notices of Request for Proposals (RFPs). Local ordinances require that these notices be placed in certain publications, just as bid requests are. Usually they appear in at least one newspaper serving the community involved.

A representative of the cable company that wants to be considered for a franchise contacts the municipal authorities and gets a copy of the RFP. The document spells out the questions that the municipality wants the cable companies to answer before it awards the franchise. Often the municipality wants to know:

- what rates the company wants to charge, both for basic and pay services
- what programs the cable company wants to show
- how many channels the cable company will have
- financial resources of the cable company (Does it have enough money to do what it promises?)
- local access and local origination plans
- plans for training of local residents in the use of cable equipment
- conditions of service (hours cable will be available, special terms the municipality wants to impose)
- job opportunities (number of openings)

In return for granting the franchise (which allows the cable system to service the area exclusively), the municipality gets a franchise fee. Specific terms vary from community to community and are negotiated as part of the award process.

Franchises are usually written with a specific expiration date. The expiration date protects both the cable company and the municipality. From the company's point of view, it has the territory exclusively for the period guaranteed by the franchise. If the company can increase sales to the residents, getting more people to take cable (and especially, premium services), they'll have greater income. No other cable company can compete with them.

The expiration date protects the municipalities, too. When the franchise expires, the cable company will have to renegotiate the terms, and will almost certainly face competition from other bidders, especially if the system has been profitable. At the same time, the stiff terms of the franchise give the municipality a hold over the cable companies. The municipality gets revenue income yearly. And the cable companies *must* perform satisfactorily.

When a franchise is renegotiated or transferred, the municipality may try to change its terms. For instance, United Artists Entertainment originally held the cable franchise for Los Angeles. In late 1991, UAE was absorbed by Tele-Communications, Inc. The City of Los Angeles had to approve the transfer of UAE's franchise—and its 90,000 subscribers—to TCI.

The Los Angeles Department of Telecommunications recommended that the city okay the switch—but only if customer-service standards were tightened. TCI, said the Department, should respond to a phone caller within 30 seconds no less than 95 percent of the time, should provide a credit for a month's worth of basic service any time that TCI missed a rescheduled appointment, and should credit a subscriber whenever an outage or outages of four or more consecutive hours in a 24-hour period affected the customer's cable service.

FRANCHISING AND REGULATION

Cable regulation goes back to 1965, when the Federal Communications Commission (FCC) adopted its first rules regulating cable television. The FCC claimed jurisdiction over cable, based on the effects that cable television had on the system of local broadcast television established by the FCC.

During the years that followed, the FCC set standards for the franchising of cable systems by local governments. But the 1975 regulations on premium cable service that FCC tried to adopt in its effort to protect local broadcasting from cable television were overturned by the courts.

In 1984, Congress finally gave the FCC explicit authority over cable. The Cable Communications Policy Act of 1984 established national, uniform procedures for franchising and renewals to assure that cable systems were responsive to the needs and interests of their communities. Under the new law and subsequent FCC rules, most cable systems were freed from rate regulation of basic services—beginning in January, 1987. Franchise fees that cities could charge were capped at 5 percent of the cable system's gross revenues from basic service.

By 1991, however, the U.S. Senate was considering new major legislation that would change the rules substantially. One version of the controversial bill would have reregulated cable rates and curtailed the number of cable subscribers served by the same multiple-system operator.

U.S. Senator John C. Danforth, who introduced the Cable Consumer Protection Act of 1991, said the primary goals of the proposed legislation were to restore public authority over cable rates in areas with a cable monopoly, to spur competition and diversity by opening up programming services to broader distribution, and to require that minimum customer service and technical standards be set.

The bill also contained important new provisions to prevent franchising authorities from unreasonably refusing to award additional franchises, or to impose unreasonable time limits on the building of competing cable systems. Another provision—opposed by both the cable companies and the Motion Picture Association of America—would have required cable companies to pay fees to retransmit the programs of over-the-air television stations.

Job Opportunities

The effect of the proposed law on franchising and on franchising jobs is uncertain. In the days when cable was expanding rapidly, the National Cable Television Association called franchising a "fast-paced area of cable, loaded with opportunity for the creative and energetic." By 1992, however, NCTA didn't even list franchising as a job category in its booklet, "Careers in Cable Television."

Nevertheless, people *do* work with franchises, and the legal concerns they bring. Often, a multiple-system operator has a director of government affairs. Such a person works with public officials and regulatory agencies to provide information about the cable company, monitors matters dealing with regulations and legislation, and represents the cable company at hearings.

Franchising and regulation have become so complex that anyone working in the field needs an extensive background and experience. One such consultant is Dr. Barry M. Orton, professor of telecommunications at the University of Wisconsin-Madison. Under the UW-Madison Department of Communication Programs, Dr. Orton and his staff offer telecommunications consulting services to Wisconsin municipalities. They work with local governments on cable-related issues, including original franchising, renewal, performance evaluations, and legal disputes.

Trade publications such as *Cable World* sometimes carry ads for franchising jobs. For instance, the City of Milwaukee advertised for a cable franchise officer. The job involved administering the provisions of the city's cable communications ordinance and the 15-year franchise agreement for the City of Milwaukee.

The candidate selected was also required to review financial data to ensure that the cable company was complying with the franchise agreement, to analyze reports from the cable operator, and to keep Milwaukee's mayor and city council up-to-date on cable matters.

To qualify for the job, which had a salary range of $37,837 to $52,696 (plus benefits), applicants needed five years of management experience with a public cable television regulatory authority, and a bachelor's degree in communications, economics, or public administration.

CONSTRUCTION JOBS

What kinds of opportunities lie in cable construction? Where are the jobs, and what do you have to know to get them?

With close to two-thirds of U.S. homes already subscribing to cable, the dramatic growth of new systems that was common in the early 1980s has slowed down. Cable construction still continues, however.

Some of this construction represents new plant, as farmlands are supplanted by suburban housing developments. Other companies may be maintaining or replacing cable in existing systems. Because of technical improvements, new fiber-optic cable is often used as the replacement. A trunk system with fiber-optic cable creates stability, so there isn't a need for as many amplifiers in the line. That means fewer chances for cable to fail, and better reception for customers.

Either way, there are cable construction jobs, and plenty of them. Sometimes cable operators do the job themselves, using employees, especially if they're only connecting a few miles of plant. Frequently, however, MSOs with substantial new or rebuild plans hire experienced outside contractors. As one construction executive puts it, "Sometimes cable television companies call us to fill in around the edges."

A Major Construction Contractor

One such contractor firm is Excalibur Cable Communications, with headquarters in Fairfax, Virginia, just outside Washington, D.C. The company's experience includes over 1.25 million residential installations, over 3,000 miles of underground construction, over 1,000 miles of aerial construction, and over 2,500 miles of splicing and activating cable. Excalibur's workers have wired more than 300,000 apartments, townhouses, and commercial buildings. They've made over 45,000 service and line tech calls.

Excalibur provides a number of services to the cable industry, including drafting and design; residential installation; pre- and post-wiring; underground construction, rebuilds, commercial installation, aerial construction, and aerial maintenance.

The company works extensively in the Washington, D.C. area. However, recent job sites have included California, Nevada, Ohio, Kentucky, North Carolina, and New York—wherever the jobs are, says Charles E. Kirtley, Excalibur's director of marketing and new development. "If we migrate to a new state, we'll often pick up subcontractors for the work," Kirtley explains. "At other times, we'll run ads in the major newspapers. We'll look for past experience."

About 35 percent of his company's jobs are in construction (including rebuilding systems), about 45 percent are installations, 5 percent are in system design, and 5 percent are in audits. Excalibur "may or may not" check references for construction jobs; virtually always checks them for installers. In a few cases, the company is asked to bond workers.

Kirtley, a former contractor himself, says he's hired and trained over 200 cable television workers during his career. When he hires, the quality he looks for most, he says, is the ability to learn.

Construction Crews

Cable construction jobs are more-or-less standard throughout the industry. "Laborers are the people who are down in the trenches for underground construction," Kirtley says. "They come along behind the trenching machines, doing backfill and restoration. If you're putting in overhead cable, the laborer is the ground person who helps out, making sure materials are sent up to the aerial crew. Pay for laborers starts at minimum wage and goes to $7–$9 an hour."

Qualifications? Good work ethics, according to Kirtley. These include showing up on time, good attendance, and good on-the-job performance. Laborers who are "high producers" eventually move up to be machine operators, Kirtley says.

Machine operators are paid from $10–$25 an hour, depending on their experience, the geographic area of the job site, and the difficulty of the task. They operate backhoes, trenchers, plowing machines, jackhammers, air compressors, and similar equipment.

Pay for foremen and supervisors ranges from $15–$30 an hour, depending on the scope of the job, Kirtley says. Excellent references and at least five years' previous experience are needed. "Such people are knowledgeable about the trades, and can oversee the job site from beginning to end. Some may be salaried, and are considered as a part of management."

In a cable company like Excalibur, jobs don't always fit the 9-to-5 time slot when you're laying cable. Kirtley says that normal working hours for construction crews are built around a 40-hour, five-day workweek, since construction can be preplanned. "But installers are out there to make customers happy," he says. "You can expect to be working nights and weekends, or whenever the customer can make an appointment. You may see anything from 4-day 10-hour-a-day work to 7-days-a-week configuration. There's a lot of shift work."

When Excalibur hires installers, the company looks primarily for men and women ages 20 through 35 who are "clean-cut in appearance," Kirtley says.

Franchising, Construction, Financing 33

> An installer has to be a combination of a laborer and a diplomat. Those are two tough qualities to find in the same person.
>
> Installers have to be good with their hands. They need to be physically able to climb ladders and phone poles. They have to be technically capable of understanding what is going on inside the television set. But they also need to be able to go into a home and hold a face-to-face conversation with a subscriber. When we visit customers' homes and keep appointments, we basically represent our clients—the local cable television companies.

Cable television installers work in a variety of neighborhoods—as many as four or five a day. They need to be able to get along with people from different backgrounds. As Kirtley puts it, "One day, you're working in a meticulously kept, upper-echelon housing area; the next job may put you in a neighborhood that's drastically different."

Pay for installers is often determined by the geographical area in which they're working. Through market surveys, many cable companies try to keep salaries competitive. Persons with virtually no experience working in brand-new neighborhoods may average $6–$9 per hour. In neighborhoods where there's a high cost of living, hourly pay is usually in the $8 or $9 range. Excaliber pays benefits to its own employees; if installers are working for firms that have subcontracted to Excalibur, the subs generally provide benefits. Employees use Excalibur-owned vehicles, and the subs often provide installers with vehicles to drive.

Auditors are hired by Excalibur when a cable television system asks the company to do an audit. "Many systems audit every five to ten years," says Kirtley. "They're concerned as to whether all the people hooked up to cable are really paying for the service."

At Excalibur, it isn't always necessary to have a high school diploma for entry-level jobs. "You need to be able to read and write, however," Kirtley says. "If you're an installer, you'll have to document your doings daily. For record-keeping, you'll need to put down just what was accomplished at each home."

For technical jobs, Excalibur likes to see a background with a trade school or program that offers certification classes in technology like the splicing of fiber. Read the trades to find listings of companies offering the classes, or look in publications like *Cable Yellow Pages,* a comprehensive, convenient source of industry telephone numbers and valuable vendor information.

CABLE FINANCE JOBS

The broad area of cable finance has changed a great deal over the years. Media analyst Paul Kagan has reported that in 1985, there were approximately $6 billion worth of cable transactions. In 1988, the peak year for sales of cable systems, Kagan reported approximately $14 billion. That figure dropped to approximately $8 billion in 1991.

Initially, cable companies were financing new construction. Construction costs were expensive, especially for underground cable, which costs nearly twice as much as aerial installation because there's more risk and difficulty in digging around utility lines and in excavating rock. Today, cable systems still borrow money to upgrade systems—perhaps, to install fiber-optic cable. But much of today's lending finances working capital for expenditures, or for future acquisitions.

The trend towards consolidation of cable systems began in the 1980s, as the market began to mature. Many of the smaller systems operators sold out, so that by 1991, the top 20 companies controlled nearly 80 percent of the U.S. cable market.

That year, the total financing of cable systems throughout the United States represented approximately $40 billion of investment according to John Shemancik, executive vice-president of the media business division, Heller Financial, Inc.—a company that provides financial services to cable.

"Looking at the overall picture, media is a big business," Shemancik says. "Expenditures for all media (including cable) are the 6th largest component of the U.S. gross national product. Consequently, media lending has become a specialized industry. Financial institutions like Heller; U.S., Canadian, and foreign commercial banks; and insurance companies lend money to cable. But they also lend money to broadcast television, radio, and other media entities. As well, a number of cable companies use public trading on the New York Stock Exchange."

Within Heller, Shemancik heads a 26-person staff that concentrates on the media business. At Heller, of course, *media* encompasses much more than merely cable.

Heller has organized its staff around functional areas: marketing, underwriting, and portfolio management. Some of his staff are marketing-oriented. Generally they are senior individuals, with 10–20 years experience. "They maintain industry contacts," Shemancik says. "They access transactions—that is, find sources of business. And they financially structure loans."

Underwriters look carefully at proposed transactions, making sure that financial statements of the organizations that want to borrow money accurately reflect a company's business. Underwriters also check out just where a particular company "fits" within the context of its industry.

Once a loan has been closed and money has been funded, Shemancik's portfolio management team is responsible for that particular account. Portfolio management tracks the financial and business trends of various clients, as well as the portfolio overall.

Responsibilities for human resources and legal matters come under still another functional area. Shemancik's media division also includes a financial staff managed by his divisional controller.

About 90 percent of Shemancik's staff have graduate degrees—most often an MBA. But Heller does hire men and women who've just graduated from college for the company's general financial management development program. Heller interviews on campus, Shemancik says, and takes four to six new-hires a year. Starting salaries range from $30,000 to $40,000 a year depending on the level of experience and type of education. During a two-year period, trainees rotate through the company's functional areas, including Shemancik's media division. Some have later joined Shemancik's staff.

Heller also hires persons with several years of industry experience, Shemancik says. Usually, the company looks for a background in lending or accounting, as well as the undergraduate degree. Salaries may range from $30,000 to $40,000 depending on experience. For those with an MBA and industry experience, salaries may begin at $40,000 or $50,000.

"Computer skills are an absolute must," Shemancik says. "We do a lot of financial scenarios, including projections over a 5-year time frame. You've got to know how to run a spreadsheet, and how to project the basic dynamics of various businesses."

Overall, cable lending represents an attractive career opportunity for individuals willing to invest the time to specialize in their financial and educational backgrounds. It is a growing opportunity with changing technology and ever-changing needs that will continue to offer future potential.

BROADCAST CABLE FINANCIAL MANAGEMENT ASSOCIATION

Many men and women who work in financial management and related areas for cable *operating* systems belong to Broadcast Cable Financial Management Association (BCFM), a nonprofit professional association. BCFM members receive information tracking financial trends, regulation changes, and business theory, targeted at the broadcast cable industry. The association's six-times-a-year journal for members covers practical ideas and successful case histories. An annual conference, (cosponsored, in 1992, with *Electronic Media* magazine) includes forums and panels on cable-related issues.

CHAPTER 5

TECHNICAL CAREERS

Job prospects in cable are bright for those interested in technical positions. Cable system operators throughout the country feel there's a real shortage of qualified personnel, which may worsen as many of the older cable systems convert to newer technology.

The old lines that divided "technical" jobs from "people-oriented jobs" may be blurring, according to various cable veterans. Some, like Excalibur Cable's Charles Kirtley, believe installers need a combination of technical expertise and "people" skills, since they work in subscribers' homes and have face-to-face contact with customers. Susan Hook, general manager of WESTSTAR Communications III in Bishop, California, says one of her system's current customer service representatives spent a year as an installer; the experience gave the CSR a new (and valued) perspective on customer problems.

But for those who like electronics and technology, the future is bright. Rhonda Christenson, general manager of Continental Cablevision of Northern Cook County (Illinois), predicts "a new, blue sky for technical opportunities. Technologies coming down the line," she says, "include fiber optics and digital compression. Because cable has evolved so fast in a short amount of time, the industry is eager to acquire and retain individuals who have had the skills to jump in, and to learn quickly."

A SUBURBAN SYSTEM

Near Chicago, Continental Cablevision of Northern Cook County has 47,000 customers in five northwestern suburbs. The system carries 43 network and satellite channels and seven premium services, two of which are pay-per-view. General manager Rhonda Christenson oversees day-to-day operations.

"If a new subdivision is being built," she says, "our design technicians work out the best plan for cable. How can we extend it most easily: through the front of the streets, through backyard easements? Where will the various pedestals—boxes that house cable electronics—be located?"

Starting salaries for *design techs,* she says, are about $23,000. Eventually they can earn up to $33,000.

Technicians also start at $23,000, and can top out around $35,000, Christenson says.

Installers, usually hired at entry level, learn how to wire houses and put in additional outlets. They trouble-shoot problems, whenever possible.

At her system, *service technicians* respond to individual calls from subscribers, making repairs.

Line maintenance technicians, who help keep the system up and running, are responsible for the electronics throughout the five-suburb system. Although they're on call 24 hours a day and carry pagers, generally they work an eight-hour shift, five days a week.

"Blown fuses or other equipment failure affects our customers' service," she says. "Consequently, we focus on preventive maintenance. Our line technicians frequently balance each amplifier station, making sure the frequency of the signal that's passed on falls within the required specifications."

Overseeing the nearly 50 technicians who work outside, she says, are five *technical supervisors.*

Although a college degree isn't a job prerequisite for a line technician, Christenson looks for an electronics background, plus skills and interest, when she hires.

Like most systems, Continental has a *headend technician.* "Sometimes the position is called a *chief technician,*" she explains. "At our system, the job includes responsibility for all the satellite receiving stations and the modulator/demodulator equipment. That's the equipment which transfers signals into an RF signal that leaves the building."

Depending on the size of the system and complexity of the technology, she says, salary range for this type of job runs from the $20s to the mid-$30s. The position is usually held by the senior staff member with the greatest amount of electronic experience.

In 1992, Christenson's system began to install a fiber optics trunk system. "The new technology allows us to bypass the current amplifier cascades," she explains. "We can send a good-quality signal from start to finish—from headend to fiber node—without any kind of amplifying, so the quality of the signal at the fiber node is much more reliable.

"Customers living close to our office won't notice that much difference in the signal; customers living farther away may be able to tell the difference. But if there's an interruption in power or signal, it will affect fewer customers."

The five Chicago suburbs Continental's system serves have students in two different high school districts. Some of the students are in District 214, and are served by TCI, another cable system owned by a different cable company.

Originally, Christenson's Continental Cablevision system was constructed by Warner Amex, who gave it two-way interactive capability. Continental has put in an interconnect with TCI. Christenson explains how it works:

> High School District 214 can teach a class in one of its buildings and be linked by audio and video with the six other schools. It doesn't matter whether we serve the building or TCI does.
>
> For instance, there may not be enough students at any one of the schools to support a class in Japanese. But District 214 can have one teacher at one location. Using several different cameras, they send the signal to the other schools, in which students are watching. The students have immediate, interactive contact with their teacher. They take their tests at their home schools, and fax the results to the teacher.

A RURAL SYSTEM

Christenson's cable systems are concentrated within a relatively small, suburban area, making it easy for technicians to service customers and do repairs when necessary. A number of cable systems, however, serve viewers in rural areas. Many of these systems are older, and are considering the cost-effectiveness of changing to newer technologies. Job opportunities and responsibilities for technicians at these systems are somewhat different.

In Coalport, a community of northwest Pennsylvania, Richard Ginter is president of CPS Cable Vision Inc., and general manager of the cable system. Like others in the industry, he hadn't previously planned to get into cable. Ginter had dropped out of college and served in Vietnam. When he was discharged in 1969, he'd been trained to run computer-operated machines. His uncle, formerly a field engineer for Jerrold Electronics, had built the cable system in Coalport, and Ginter started on the ground floor.

Today, CPS Cable Vision serves eight municipalities, with about 1,920 subscribers. Most of the municipalities are small. One township has perhaps 2,200 subscribers; another borough has just under 800. The system provides 24 basic channels and three premium services: HBO, The Movie Channel, and The Disney Channel. One channel is reserved for local access. Ginter also provides FM retransmission service; customers can hook up an FM receiver to the cable system and get better sound than with a normal cable antenna.

"I service down to 10 homes per mile," Ginter says. Five full-time employees, including himself, run the system, which also has several part-time employees. Full-time employees get hospitalization benefits, ten holidays, one week vacation after the first year (two weeks after the second year), and can work up to three and four weeks of vacation, depending on length of service. The system has 90 miles of plant—that is, there are 90 miles of cable.

Ginter himself is president of the corporation, the general manager who does computer programming, and the headend maintenance technician "most of the time." Because it's a 24-hour system, one of the employees is always on call, and Ginter carries a pager at all times. Using the pager, he can determine whether the system headquarters has power or has an electrical outage.

Two installer/technicians service all customers. A typical day for one of them may include working through a list of disconnects and activations, going out to customers' homes. Part of the day is spent in trouble-shooting customer complaints—finding out whether the problem is local, regional, or systemwide.

Sometimes the technicians work together, especially on construction, rebuilding, or cable maintenance. They do all construction and installation, from the tower site at the system's one location to the customer's home and to the back of the TV sets, Ginter says. "We've just built three miles of plant to extend the system out to a new trailer court and campground."

One reason for the two-person team is safety. "There may be a broken pole, and we need traffic control," Ginter says. "One tech will be fixing the pole, and the other will be directing traffic."

They also routinely maintain amplification along the cable. "When we have cascades of amplifiers that send the signal downstream from our headquarters to the customers' homes," Ginter explains, "we know they have to be kept within certain parameters. If we can check them out, and find those that don't fall within the range, we can pull out the bad ones and put new ones in—before our customers call us to complain of signal problems."

Defective or malfunctioning amplifiers are tagged, sent outside the system for repair, and are retested by Ginter's techs before they're put back in service.

To meet FCC requirements, Ginter's system must monitor signal leakage quarterly, and must file an annual report with the government agency. He views checking signal leakage as an early warning of possible trouble for subscribers.

Ginter suggests persons looking for technical jobs in television should have hands-on experience and training with electronics. He also suggests they know how to use computers. "We already do our timecards on computers," he says. "We have a database of information. If a tech problem shows up and someone is on call, our computer system can break out the problem into bite-size areas. They know where to go."

In Ginter's rural system, techs must be capable of handling an 80- or 90-pound extension ladder, and of climbing poles. He also checks driving records. The system owns a service truck, aerial lift truck, and pole truck. Technicians drive extensively on rural roads.

Much of Ginter's on-the-job training came from his uncle. Still more came through the training program offered by the Society of Cable Television Engineers. He himself is SCTE-certified as a broadband communications technician, and is working on his "engineering" designation.

Ginter serves as secretary-treasurer of the area's SCTE chapter. "Because I'm a small system," he says, "it helps to be part of this national organization and have access to their resources."

A REGION

Mark Wilson is regional operations manager for Multimedia Inc., a diversified media company based in Greenville, North Carolina. Its cable television division has headquarters in Wichita, Kansas. The division has 100 cable systems, which serve 375,000 subscribers in Kansas, Oklahoma, Illinois, and North Carolina.

Wilson has "operations" responsibilities for the Kansas region. His Wichita system serves 95,000 subscribers. Fifteen rural systems (combined) serve 15,000 customers. The Wichita system has 37 channels: six are premium, one is pay-per-view.

Many cable systems use "installer" as the entry-level technical job, with "technician" ranked higher. In Wilson's Wichita system, installer and technician responsibilities are combined and are designated as "technician." Sixty-eight technicians in Wichita handle installation and service.

"When we hire," Wilson says, "we advertise in local newspapers. We're generally looking for someone who has a high school diploma and preferably some skills in electronics. Often those skills were learned in the military."

> We're looking for a good work ethic. Are they willing to come to work on time? Are they dependable? We are trying to downsize management, and push as much responsibility as possible to the lowest levels. We need people who like to work with others. We need people who can learn, and are willing to.
>
> We'll do all our own training. We train in basic technical and trouble-shooting skills, basic electronics, and then electronics specifically related to cable television. Increasingly, we're also training technical people in sales and customer service skills. That way, a technician can go to a house, get the customer hooked up or fix the cable system, and hopefully even sell an additional cable service.

Wilson's region provides vehicles, tools, equipment, and uniforms needed for the technicians' jobs. Entry-level salaries are $7.02 per hour, plus commissions for sales technicians. "On the average, these may provide an additional $1,500 to $2,000 per year," Wilson says. The company also provides a full range of benefits.

"We give salary increases every six months. We give an annual cost-of-living raise, and also annually give merit increases that can range up to seven percent. After five years with us, a technician could be making $8.14 to $11.41 per hour, with bonuses for weekend work."

Wilson's company exclusively promotes from within. Everyone in management at his office started at entry level. Wilson himself began as an installer, although he had a bachelor's degree in liberal arts from University of Nebraska. Since coming to work in cable, he's earned an MBA from Wichita State University. The company pays for education, Wilson says, and a number of people have earned degrees while working their way up. Some have earned master's degrees, with the company picking up the tab.

Schedules for technicians rotate, he says. The normal day runs from 8:30–5:30. Technicians report to the facility and pick up work assignments, scheduled by a computer program. Throughout the day, they keep in contact with headquarters via two-way radio, receiving additional assignments as available. They go directly home when work is completed, and don't have to return to headquarters.

Next level up from technician are *technical supervisors,* who may supervise 13 to 15 people. Starting salary range for supervisors may be $35,000 to $45,000. Supervisors' responsibilities include performance appraisals, training, dealing with or supervising contractors, some light engineering duties, and coordinating with outside agencies: utilities, municipalities, or large customers like universities, hotels, or motels.

Supervisors meet frequently with their teams, to update them on current information and to serve as liaison between the techs and the company. They spend a good deal of time in helping to train their technical teams, working not only on upgrading technical skills, but also on dealing with special projects and unusual problems.

"Now the pyramid gets narrower," as Wilson puts it. On the technical side of the region, a *system engineer* (often called a chief engineer) deals with the headend—the origination of the cable signal that goes out over the system. Usually such a person has a bachelor's degree in engineering, or certification as a cable television engineer.

Wilson himself is on the management side, he says. His responsibilities include overseeing not only the region's technical department, but the construction, engineering, materials, purchasing and handling, and support services personnel at the electronics repair facility.

Wilson foresees more emphasis on technical skills and training, as new technology is adapted. His company is considering upgrading from 35 to 77 channels, he says, using more complicated fiber optics and possibly *multiplexing.* Because competitors are providing video-type services, Wilson believes technical personnel will have to put more emphasis on customer relations.

He's working on getting all the technicians in the region certified at the technical level in the certification program run by the Society of Cable Television Engineers. He also wants supervisors to be certified by SCTE at the "Engineering" level.

SOCIETY OF CABLE TELEVISION ENGINEERS (SCTE)

There are a number of engineering societies which sponsor workshops and conventions, and which publish technical material. One of the most important of these is the Society of Cable Television Engineers (SCTE).

Membership in SCTE is open to anyone involved with cable television, radio and television broadcasting, or other allied communications fields. A majority of members are not engineers with degrees. Instead, they are individuals who want to increase their working knowledge of communications technologies by participating in the society. Student memberships are also available at a reduced fee.

Members receive a free subscription to *Communications Technology,* SCTE's official journal; a free subscription to *CED* magazine; a free subscription to *Fiberoptic Products News;* and a discount subscription to *Multichannel News,* the cable industry's weekly newspaper.

SCTE Certification

SCTE has developed the Broadband Communications Technician/Engineer (BCT/E) Professional Designation Certification Program, which covers seven areas of cable technology. This program is a peer recognition program, and not a license. The program, begun in 1985, has grown until in 1990, more than 1,700 candidates were enrolled.

A candidate who wants to qualify as a *Broadband Communications Technician* must have two years of technical experience in cable television, three professional references. He or she must pass written examinations. SCTE national membership also is required.

Candidates who want to qualify as *Broadband Communications Engineers* need five years of experience in the broadband communications industry. In at least two of those years, they must have contributed responsibly to a significant technical function.

Candidates for the "Engineer" designation also need formal electronics training or its equivalent. They must have completed some form of organized electronics study: military technical training, technical or vocational school, community college or university courses. Applicants also need three professional references, from persons familiar with their work. Candidates also must pass written examinations. SCTE national membership is required.

Examinations are given in the following categories:

- Signal Processing Centers

- Video and Audio Signals and Systems
- Transportation Systems
- Distribution Systems
- Data Networking and Architecture
- Terminal Devices
- Engineering Management and Professionalism

Because technology is changing so rapidly in cable television, SCTE has a mandatory recertification program. Certified technicians and engineers must maintain their status by being recertified every three years—either through retesting, or the accumulation of the required number of recertification units. Units can be earned through a combination of SCTE membership and SCTE activities, industry-related courses, speaking or presenting a paper at a national, regional, local, or chapter conference, or by publication of an article in an industry trade publication.

SCTE also offers a similar *Installer* Certification Program for installers and installer/technicians. Candidates must pass a written exam, as well as two practical exams: proper drop cable fitting preparation and installation, and signal level meter reading.

Technical seminars sponsored by SCTE offer beginning and advanced three-day hands-on technical training at two levels. An initial "Technology for Technicians" seminar is designed for installer/technicians, service technicians, and their field supervisors. A similar, advanced seminar is designed for broadband industry maintenance technicians, chief technicians, and system engineers. Both offer a combination of comprehensive technical theory and actual "hands-on" experience in laboratory settings.

SCTE's annual meeting is open to persons from all levels of the cable television and related businesses, including all levels of nontechnical personnel. Student SCTE members can attend.

The meeting starts with a day-long engineering conference on new technologies, FCC compliance, and technical management. Cable-Tec Expo, which follows, features $2^1/_2$-days of technical workshops: test equipment, system tests and measurements, and fiber optics were recent topics. Also on display: exhibits of cable hardware from manufacturers and distributors.

In addition, the society has presented more than 75 national technical programs in cities across the United States.

For more information, contact the Society of Cable Television Engineers, SCTE, 669 Exton Commons, Exton, Pa. 19341.

NATIONAL CABLE TRAINING INSTITUTE (NCTI)

For more than 25 years, many cable television system operators, contractors and cable industry vendors have used NCTI's correspondence self-study courses as one way to train their employees to construct and operate cable television systems.

This home-study school offers a five-level career path that covers all areas of the cable system: from the customer's television set to the headend. Courses include: Installer, Installer Technician, Service Technician, System Technician, and Advanced Technician.

Most NCTI students are already working in the cable television industry, or for companies that have an interest in cable television. It's possible, however, for an individual not connected with cable to enroll.

At the Installer level, for instance, students learn how to perform aerial, underground, and interior cable television installations. The course begins with a comprehensive overview of a cable system—giving students a working knowledge of the system, from the signal sources to the customer's television set.

In addition, the course includes lessons on using and maintaining tools; equipment; and techniques for pole climbing, cable routing, and installing terminal devices, convertors, and decoders. Students also learn how to tune TV sets, operate a signal level meter, and troubleshoot TV signal quality. Some lessons in the series deal with customer relations, and with alternatives customers can choose: pay-per-view, digital audio, and interactive video.

Courses consist of individual lessons, which progress in difficulty. Most are technical. Others combine technology with management skills. For example, one lesson in the Chief Tech Update series teaches

48 *Opportunities in Cable Television Careers*

students about digital communications concepts, modems, protocols and error detection. Another lesson teaches students how to detect leakage, manage outages, reduce theft, and set up a preventive maintenance system.

NCTI also has courses on special topics. They include material on CATV system overview, two levels of courses for broadband RF technicians, and a course on CATV fiber optics.

Because virtually all older cable systems were built with coaxial cable, the fiber optics self-study course discusses the new technology in depth. Students learn about fiber concepts—from *transmission* and *attenuation* to *bandwidth* and *dispersion*. They learn about fiber applications—from *cabling basics* and *types of lasers*, to *amplifiers* and *splicing*. Students also review fiber architectures, modulation techniques, RF interfaces, components, testing and monitoring, construction and maintenance.

An in-depth special course on CATV Technology for Nontechnical Personnel teaches students about the history and development of cable television, where program signals come from, how they are processed and travel through the system, how an installation is performed, and how television sets produce pictures.

Other special courses cover customer relations, and television/video production. The latter includes lessons on studio and portable production; videotape recording, visuals, and scriptwriting; and videotape editing.

NCTI also offers on-site seminars, including one covering the dynamics of supervision, for cable company supervisors and managers who have had little or no formal management training, or for experienced managers who want a refresher course in management skills. Another seminar, offered at dates and locations around the country, teaches cable managers how to comply with regulations of the Occupational Safety and Health Administration (OSHA).

For information, contact the National Cable Television Institute, P.O. Box 27277, Denver, Co. 80227.

CHAPTER 6

SALES AND MARKETING

Marketing (broadly defined as the selling of cable to customers and all the activities that go with it) is extremely important to cable systems. Unlike traditional broadcasting, cable companies get a substantial part of their income from subscriber revenue—the fees they charge their customers.

A cable system that's wiring a new development or subdivision will begin to sell its services shortly after construction has been completed and the cable system has been activated. Such a system has a fairly close estimate of how many subscribers it hopes to obtain from a given geographical area long before a sales representative telephones you or rings your doorbell.

But marketing—crucial to the bottom line of cable systems—must continually be done, even with subscribers who already have signed up for cable. Since one out of every five U.S. families moves each year, a certain amount of cable marketing done by a cable system targets families who have moved into the geographical area it services.

An even greater amount of cable marketing effort is aimed at keeping subscribers happy, and at getting them to increase the use they make of cable. Since cable is a tiered service—that is, a cable system charges more money to subscribers to buy additional levels of cable service above the fee for basic cable—it's to the system's advantage to persuade viewers to sign up for more premium channels. Pay-per-view, in which subscribers pay extra fees for a single programming event, such as a sports game or a concert—also represents another potential source of revenue for the cable system.

MARKETING JOBS

In new market areas or subdivisions, where systems have just been installed, many cable marketing jobs involve door-to-door selling. There are remarket opportunities also, because families frequently move. If a subscriber disconnects and leaves town, a salesperson will phone the newly arrived family or stop by, attempting to have them sign up for cable.

Many cable systems hire employees directly to do the door-to-door calling. Others contract with outside providers, who supply door-to-door sales personnel.

Cable systems, of course, don't stop marketing once door-to-door salespersons have visited a "new build" area. Many cable marketing executives believe that marketing becomes more significant as cable systems mature, and the emphasis changes to customer relations.

Retaining current subscribers while increasing revenue is the goal, of course, of cable marketing. Techniques such as bill stuffers, television spots, telephone sales, and follow-up calls help.

Marketing "how-to's" and successful marketing experiences are described in the trade publications. Each tracks down specific tips from cable systems and passes them along.

Customer Service Representatives

Customer service representatives (CSRs) are an important part of cable marketing. A CSR may handle more than a hundred phone calls a day. Many are from persons whose sets aren't working properly or who want to change services. Customer service representatives are an important factor in reducing *churn,* the industry term for disconnecting or downgrading services.

It's not always easy to break in. "Our marketing department has a very low turnover rate," warns Theresa Wright, employment specialist in the Human Resources Department of Media General Cable of Fairfax, a Virginia cable company.

"Most entry-level marketing jobs are clerical or administrative. Usually these positions are filled by recent college graduates who generally major in marketing or communications. Starting salaries are mid-high teens." Media General Cable, like many other cable systems, likes to promote from within the organization, if it possibly can.

Local Jobs

Local cable systems also have traditional office jobs which use skilled workers, and which may offer you a chance to break into cable. They need persons to do accounts payable, accounts receivable, general ledger, personnel (often called human resources), and other traditional business functions you'd find in any medium-size business.

MARKETING AT TWO CABLE SYSTEMS

At Warner Cable in Houston, Texas, Marc Herz is vice-president of sales and marketing. The system is owned by Warner Communications, a Time Warner subsidiary, and serves parts of the Houston community.

Herz, who's been involved with cable for ten years, is responsible for the sales function. He manages the sales force and the telemarketing force, and supports other departments' efforts to sell cable service. He devises and implements the marketing plan, and interacts with vendors and suppliers of cable television programming services. Herz, who has a master's degree in business administration and an undergraduate degree in accounting, also works on business projections and forecasts.

For door-to-door sales representatives, Herz looks for "bright, energetic individuals with good communication skills. They must be self-motivated," he says. "They must come across as presentable to the public. They create the company image, so they must be courteous and respectful to subscribers."

At Warner Cable of Houston, direct sales are commission positions. Sales representatives generate their own leads or follow up leads given to them. They need to have their own car, insurance, and a driver's license. They receive one to three weeks of company-paid training initially, and additional on-the-job training during their career with the company.

"We teach them what to say to the customer when they knock on the door," Herz explains. "We want them to pay attention to the quality of their interaction with the customer."

The cable system uses a variety of compensation methods, Herz says. One plan uses a salary base (new-hires start at $12,000 and earn increases up to approximately $18,000). Another plan is commission only, with the average commission for a sales representative ranging from $18,000 to $25,000.

Sales positions are typically full-time, but telemarketing positions (which start at $6 an hour) are usually part-time, mostly for about 20 hours a week.

Jobs for customer service representatives (CSRs) are "highly sales-related," Herz says.

A college degree isn't a necessary prerequisite, he points out, and although "we tend to like people who have completed high school, it's not always a requirement." Previous sales experience, however, is a plus.

In fact, Herz feels some college graduates are unrealistic in their expectations. "They tend to come in and expect a great deal of responsibility with little experience," he says. "It's not realistic for someone just out of college to get a management position right away. There are a lot of people already working in cable who deserve to be promoted. We like to give opportunities to those who are already working for us.

"Even if you have a master's degree, you ought to be willing to start as a telemarketer, if that's the job I want filled," he points out.

Herz says he'll give anyone who is energetic and wants to work hard an opportunity, if a job is available, even if they don't have previous experience. His advice to those who want to break into cable: "Even if the position doesn't suit your skills, start somewhere. Realize that our industry is constantly changing . . . is in transition. If you have a two- or three-year time horizon, you'll advance, but you must be prepared to make that commitment. You need a level of maturity sufficient to understand that rewards for your performance may be put off until tomorrow."

During job interviews, he's turned off by candidates who don't present themselves well. "Someone who doesn't speak clearly, doesn't use proper English, has a resume or cover letter that's unprofessional or looks unsophisticated—or who doesn't know what to expect in the workplace—just doesn't come across."

Candidates should be looking for "fit," he says—"what's the organizational culture, and whether or not they feel comfortable within it.

"You need to ask: 'How often is my performance reviewed? How much supervision will I be given? On what basis will my performance be appraised? What's the disciplinary process like? How would you describe this organization—relaxed or formal? loose or controlled?' "

Herz says those he hires need to demonstrate not only proficiency in their assigned tasks, but also a willingness to work hard and harder. "They need to show they're ready for additional responsibility," he says. "By demonstrating their willingness to undertake more, they're showing they are ready for promotion to the next level. They may advance to be supervisors responsible for several people who are doing the work they'd previously done."

Cable marketing objectives have changed considerably since cable's early days. Most cable systems are moving towards maturity, so sales and marketing opportunities are in "remarket."

Philip A. Roter, vice-president of sales and marketing for Cable TV Montgomery, in the Washington, D.C. suburban area, says systems still need to add basic cable subscribers—"to convince people who've previously resisted—who've said 'no' to cable" the first or the tenth time they were asked.

"We need to 'take away' the reasons they haven't subscribed to cable," he explains. "That means we need to use narrowcasting and segmented marketing, as well as to provide new programming, so customers have more options."

In the future, he feels, cable marketers will need to target special interest groups. But it's not easy. "When you have 67 channels running

24 hours a day, as we do," he says, "it's going to be more important than ever to communicate to potential subscribers just how cable programming can meet their special needs."

Roter also believes Pay Per View (PPV), and advertising sales will become increasingly important as revenue factors to cable systems. "We'll be focusing more on movies," he says, "as cable becomes the alternative to the video stores."

Solid sales and marketing—as Roter explains—have been the key to cable's success. Sales and marketing skills are needed even more, Roter feels, as cable's future changes.

> As an industry, we're getting more analytical. We need people in sales and marketing who bring us experience—either through courses they've taken, or through the work they've done before they come into cable. Cable sales and marketing will need people who understand segmenting, demographics, psychographic analysis, and life-style. That's pretty sophisticated. But that's where the jobs will be.
>
> Sales and marketing are good areas for breaking into cable. In our cable system, which has 175,000 subscribers in the Washington, D.C. suburban area, we have 212 people working in sales and marketing. About 120 of them are customer service representatives and direct sales reps.

Although the system offers customer service round-the-clock, CSRs work an 8-hour shift, earning between $8 and $12 an hour.

Though Cable TV Montgomery welcomes a background in marketing or communications courses, the system doesn't specifically require them. Instead, the company screens and hires primarily for "people skills."

"We look for people who are energetic, enthusiastic, and enjoy what they do," Roter says. "They have the ability to 'smile' when they talk to customers. They communicate clearly."

CSRs are asked to meet performance standards. "We pay attention to average hold times," Roter says. "We don't want customers on hold for longer than 30 seconds. We watch abandonment rates—people who hang up without having someone take care of their questions. We don't want to lose more than a small percentage of calls."

At Cable TV Montgomery, 70 direct sales reps call on nonsubscribers. Their target is to "pass" the market $3^1/_2$ times a year—that is, to contact a noncable household between three and four times in a 12-month period.

"A direct sale is the most effective sale," says Roter. "If we can get in front of a prospective customer and spend 15 to 20 minutes explaining the scope of our product, we then have an informed customer at the point of installation. Our customers will use the product correctly and find value in it."

"Usually," Roter says, "direct sales reps will be successful with approximately 15 percent of the nonsubscribers they visit. It's a job in which you've got to cope with a fair amount of rejection," he warns. "But a good direct sales rep in our company will earn about $29,000 a year on average. Some can make up to $75,000 a year."

To reach people who won't buy cable through direct sales, he says, Cable TV Montgomery develops and sends direct mail pieces. Some direct mail is blanket-oriented, such as a recent 6-week campaign for HBO and Showtime. "The same mailing piece was sent in Illinois as in our Washington, D.C. suburban area," he explains.

Often, though, he prefers specific direct mail. "Last month we did a bilingual piece with a Spanish lead. We sent it to Hispanic households. Another mailer—featuring Wimbledon tennis on HBO and the U.S. Open on USA Network—was sent only to members of local tennis clubs. Message targeting appeals to specific interests. It helps us get cable in front of hard-core nonsubscribers, and lets us enlighten them about what we can provide."

Other direct mail pieces are aimed at move-ins. "We send invitations to new households to become subscribers," Roter says. "We buy lists of new phone numbers and give them to telemarketers who phone 30 days after the move, offering 'move-in' special rates."

Roter hires outside marketing research companies "as needed," and uses Cable TV Montgomery's own marketing analyst to work directly with them. "He holds focus groups every month to get an indication of how subscribers feel," Roter says. "He tracks trends in industry and business."

Unlike CSRs, who work on an hourly basis (with benefits) or direct sales reps, whose pay is tied to commissions on sales, at Cable TV Montgomery, the marketing analyst and marketing coordinator are salaried. "Analysts generally earn between $20,000 and $30,000 a year," Roter explains. Such a position requires cable experience or prior research, hopefully within a media-related industry.

"A marketing coordinator usually makes between $30,000 and $50,000, and often has an ad agency background. However, our marketing coordinator came in at entry level and performed so well that she was promoted within our system. Now she's a major force in helping to make marketing decisions. She handles direct mail, including all forms of collateral materials and virtually all communications that go out to subscribers."

Cable TV Montgomery also has a PPV director to run its rapidly growing $4 million annual pay-per-view business. "In the mid-eighties, there was no such thing as PPV," Roter says bluntly. "By mid-'92, about half of U.S. cable systems were in the PPV business. Now we have a $60,000 director of PPV who has four years' experience in pay-per-view."

"New technologies are opening up opportunities!"

In 1992, Roter says, his system offered 67 channels. Within that lineup, seven premium services and three full-time pay-per-view channels were available to subscribers. One of those channels served as a "barker"—running 24 hours a day to promote pay-per-view offerings.

Down the line, he predicts, are technologies like digital compression: the ability to squeeze the signals together electronically so a cable system can offer increased numbers of channels, yet still use the same amount of "space."

"Digital compression will give a cable operator the ability to turn one video channel into four or six video channels without having to incur tremendous costs," he says. "Before the turn of the century, we'll be able to put 150 to 200 channels into your home! But without digital compression, we couldn't increase our channel capacity without completely rebuilding the cable system."

When that time comes, Roter says, sales and marketing opportunities will be expanding rapidly.

"How can you communicate what 200 channels can mean to a customer?" he asks. "You can't do it through a cable guide.

"There'll probably be 50 channels of PPV for video-on-demand. At any one time, subscribers will be able to choose from 50 feature films."

Systems that will use the technology, he predicts, can provide niche programming. "If I'm a DePaul Blue Demon fan, they can put together and sell a package that lets me see every one of DePaul's basketball games. If I'm a Phoenix Cardinal fan, I'll be able to buy a package of their games that hasn't previously been available.

"Cable will need even more marketing people and customer service reps to sell and service cable offerings."

For persons who want to succeed in cable marketing, Roter says the ability to be flexible—to be comfortable with rapid change as the industry evolves—is crucial.

Roter's own background illustrates what he means.

After he graduated from DePaul University with a degree in business management, he began working for TelePrompTer as a direct sales rep. He spent seven years with the company.

"I didn't have the same job for more than 18 months in those seven years," he recalls. "I moved from smaller to larger to increasingly larger responsibilities."

When the company was sold, Roter was recruited through a job search firm and chose to work for Harte Hanks, a cable MSO. He started there as the regional director of sales and marketing, responsible for increasing revenues for five Texas cable systems. By the time that company was sold five years later, he'd become corporate vice-president in marketing and programming, responsible for 14 cable systems in seven states.

Again, a recruiter contacted him—and Roter became vice-president of sales and marketing for Cable TV Montgomery, near Washington, D.C. In less than three years after he started, he'd increased cable penetration from 37 percent to 58 percent.

"If you want a career in cable sales and marketing," he says, "you've got to 'grow the business.' Your job is to increase the revenue, thereby increasing the overall value of the cable system."

Although Roter doesn't downplay the value of a college degree, he says customer services, direct sales, telemarketing, and possibly clerical work in marketing are good entry-level jobs. "They'll give you the experience you need," he points out. "They're front-line jobs. That's where you connect with the reality of cable. Don't think you can come out of college and start directly in marketing management. You need that customer-focused base of experience."

MARKETING A NETWORK

ESPN, America's largest cable network, televises sporting events and sports news, information and life-style programs. The 24-hour network televises more than five thousand live and/or original hours of sports programming a year.

Event programming includes the NFL; major league baseball; college football, including CFA games; NCAA basketball; men's and women's pro tennis, including the Australian and French Opens and the Davis Cup; PGA Tour, LPGA, and Senior PGA Tour golf; NASCAR, CART, and Formula 1 auto racing; thoroughbred and harness racing; Top Rank Boxing; men's and women's professional bowling; and World Cup skiing.

Special events or series include the America's Cup; NFL Draft, Baseball Hall of Fame Induction Ceremonies; Earthwinds Balloon flight; NCAA College World Series; Expedition Earth, an adventure series with a focus on environmental issues; and more.

ESPN operates three international lines of business: a 24-hour satellite network feed to cable systems in Latin America and the Pacific Rim; syndication of individual programs and series to more than 250 broadcast, satellite, and cable outlets worldwide; and participation in the management of co-venture sports networks in Europe (The European Sports Network) and Japan (Japan Sports Channel). Selected ESPN programming is available in more than 70 countries.

ESPN is America's largest cable network. Its 24-hour programming can be seen in all 50 states and the U.S. territories and possessions of Guam, Puerto Rico, and the Virgin Islands. By April, 1992, ESPN reached 59.6 million American households through more than 24,700 affiliates.

Selling ESPN to U.S. cable systems is the responsibility of George Bodenheimer, the network's vice-president for national affiliate sales and marketing, and his 17-person domestic sales staff. (A separate ESPN group handles international affiliate sales.)

Bodenheimer explains the job this way:

> Affiliate sales are very different from ad sales. Our affiliate account executives are responsible for managing all aspects of the relationship between the ESPN network and its affiliated cable systems.
>
> They coordinate efforts with ESPN's corporate office and the cable systems to market and promote ESPN. That encompasses a lot. Affiliate account executives can be doing everything from helping line up a guest speaker at the local Rotary Club meeting, to entertaining a cable system's customers at sporting events, to helping a cable sell local ad time, to showing the system our latest promotions.
>
> We hire for talent and performance. Gender doesn't matter; some of our best affiliate account executives are women.

Experience is necessary, Bodenheimer says, "but not necessarily cable experience. Obviously, we look for a college education, but we have history majors, communications majors, and those from other backgrounds on my staff."

Bodenheimer warns that the job takes energy and devotion:

> It's definitely not a 9 to 5 job. About 40 percent of their time is spent traveling. They don't just punch the clock. They need to keep up-to-date on many, many issues. They have to monitor the legislative process. They need to learn about new technology: digital compression, HDTV, and all the factors that affect the business of our affiliates, the cable systems. What affects them, affects us.

While the association with sports appeals to a lot of people, Bodenheimer points out that sporting events don't always occur Mondays through Fridays. "If you're going to entertain cable customers or be there yourself, you're often working on weekends," he says.

Bodenheimer says the most important ingredient for success "isn't experience, isn't education, but is your willingness and desire to update your skills and your abilities."

Reading the trades is an essential part of the job. "There's no shortcut," he warns. "You must read. You have to keep up with industry news, as well as with sports news."

Establishing relationships also is important.

> You need the ability to handle yourself in all sorts of situations. Some can be confrontational, so you've got to be thick-skinned.
>
> You also need the ability to work independently. We hire people, give them a set of goals, and largely try to let them manage their own affairs.

Although Bodenheimer hasn't hired anyone new for quite some time, he's often willing to see persons who want to get into cable. "I tell them that if cable is truly what they want, they'll find their way into the industry," he says. "There are jobs. The cable industry is growing, and receptive to newcomers."

Several persons he's referred elsewhere have managed to break into cable, he says, and they stay in touch with him. "An opportunity at ESPN could arise in the future," he says, "even if it's several years later."

"Make a contact, keep a contact, and develop it. That's my advice."

TRADE ASSOCIATION

The Cable Television Administration and Marketing Society, Inc. (CTAM) is the professional association for telecommunications professionals. Its board of directors includes representatives from some of the top cable MSOs.

Since 1975, CTAM has been promoting cable marketing topics through conferences and seminars. For additional information, contact CTAM's Executive Director c/o Cable Television Administration & Marketing Society (CTAM), 635 Slaters Lane #250, Alexandria, Va. 22314.

CHAPTER 7

ADVERTISING

In cable's early days, most revenue came from the fees subscribers paid to receive the signals. By 1970, advertising was being carried on only 57 of the 500 cable systems that were capable of originating programs, and on approximately 400 more of the 1,500 systems that could provide automated services like time, weather, news, and stock ticker reports.

That year, total cable ad revenue was estimated at $3 million. Subscriber revenue brought in $300 million. By 1992, however, advertising revenues reached $3.5 *billion*, according to the Cabletelevision Advertising Bureau (CAB), and are projected to reach $5.2 billion by 1995. Based on data from Paul Kagan Associates, Inc., CAB says the majority of advertising revenues come from network cable ($2.47 billion in 1992, $3.5 billion projected for 1995). Local or spot cable is the next largest category ($906 million in 1992, $1.4 billion projected for 1995); with regional sports and news a smaller-by-far share ($169 million in 1992, $306 million projected for 1995).

Cable revenues still lag far behind those of the broadcast networks, but are growing impressively. One major reason is that cable's technology and demographics offer advertisers the opportunity to reach young, educated, affluent households—persons presumably able to buy what the advertisers are selling.

"Cable TV Facts," a brochure updated yearly and available from the Cabletelevision Advertising Bureau, Inc., said in 1991 that cable reached 83 percent of viewers in which the head of the household had four or more years of college; 86 percent of viewers in which the household yearly income was over $40,000; and 82 percent of households with incomes of $30,000 or over in which the head of the household held a professional or managerial job.

The Cabletelevision Advertising Bureau tracks product consumption in cable households through its MRI Cable Report. It knows, for instance, that in spring 1991, 15 percent more cable households than the U.S. average went to health clubs, and 13 percent more cable households than the U.S. average spent $150 or more per week in food stores. Twenty percent more cable households than the U.S. average took more than three plane trips; 18 percent more cable households than the U.S. average bought a new imported car; and 14 percent more cable households than the U.S. average bought an answering machine.

Information like this helps advertisers and ad agencies make the decision to advertise on cable. And they have choices. Satellite-fed advertiser-supported cable networks include: America's Disability Channel, Arts & Entertainment Network, BET (Black Entertainment Television), CMT (Country Music Television), CNN, CNBC, Comedy Central, Courtroom Television Network, The Discovery Channel, E! (Entertainment Television), ESPN, The Family Channel, Galavision/ECO, Headline News, The Jukebox Network, The Learning Channel, Lifetime, Mind Extension University, MTV (Music Television), TNN (The Nashville Network), Nickelodeon/Nick at Nite, Nostalgia Television, Prevue Guide Channel, Prime Network, SportsChannel America, TBS, TNT, The Travel Channel, USA Network, VH-1 (Video Hits One), The Weather Channel. In addition, 24 regional cable networks carry advertising. Virtually all are sports-oriented, although New York 1 News offers 24-hour regional news programming.

Cable Network Profiles, updated annually by the Cabletelevision Advertising Bureau, provides up-to-date sales information on all the national advertiser-supported cable networks. It covers programming, commercial availabilities, demographics, audience research, and promotional opportunities. Included: network resources and contacts. For purchase information, contact Cabletelevision Advertising Bureau, 757 Third Avenue, New York, NY 10017.

INTERCONNECTS

Technology makes the process of placing the advertising easier.

Interconnects are becoming more and more important to cable advertisers. An interconnect exists where two or more cable systems link themselves together to distribute a commercial advertising schedule simultaneously. Interconnects help make an advertising schedule more effective. That's because advertisers can buy time on more than one system, yet need to negotiate only one contract. Often, an interconnect ties together an entire market.

COX CABLE SAN DIEGO

Two of San Diego's four cable systems are interconnected. Cox Cable has a "hard interconnect" linking Cox Cable San Diego and Southwestern Cable Television-San Diego. "We have an actual microwave interconnect," explains Moya Gollaher, general sales manager of Cox Cable. "Both systems get the same commercial feed. In contrast, in a soft interconnect, the tapes with the ads go to one of the systems, and are passed on to the other."

Using an interconnect, Gollaher says, an advertiser can go to one firm, get one bill, and buy time on multiple cable systems. Also, the ads that run can be very geographically specific to a targeted area.

Cox Cable's sales team is structured similarly to the advertising sales opportunities in broadcast television or radio. "We have a management team headed by a general sales manager for each local system," Gollaher says. "In a smaller system, you'd have one general sales manager; in a typical interconnect, each of the two systems would have its own manager; and in the very largest interconnect operations, you'd typically have a national sales manager as well."

Within the interconnect, Cox Cable account executives handle the sales side. They typically receive a salary plus commission. Cox runs about 30,000 spots a month across 13 channels. The *traffic department* is headed by a supervisor who handles order entry and billing for all the clients' schedules, Gollaher says. And personnel in *technical operations* make dubs and ensure the spot airs correctly. Cox averages 98–99 percent accuracy on its spots, Gollaher says—they appear when they are supposed to appear, with the "right" audio levels and video signals.

In Gollaher's system, which is in the "top 25" market for sales, salaries for those in sales can start at $25,000. Account executives at the top can make as much as $100,000 a year, she says. "A lot depends on the clients you're dealing with," she explains. "The 'better' the client list, the more you're going to make."

Entry-level salaries for clerical staff in "traffic" begin in the "high teens" and run through the $20s, she indicates, with supervisors and managers able to earn from $30,000 to $40,000.

"On the sales side, we like to hire people who are articulate, aggressive, ambitious, and motivated," she says. "They need to be attuned to customer needs. Cable is a relatively new medium with a lot of diversity. In order to match client needs, our sales staff often has to educate the client."

Basic selling skills—prospecting for business, the ability to make presentations and to follow up aggressively—are "musts" for success, she says. She likes to see prior experience in radio sales, since radio and cable both use the niche approach to marketing.

For jobs in traffic, Gollaher says, Cox prefers people who pay high attention to detail and are extremely accurate. On the technical side, the ability to focus on details is also required.

Women and minorities have good chances to succeed in cable ad sales, she believes. "I like to see some college," she says. "The big issue is: how articulate are you? We make a lot of written presentations, since we often serve as marketing consultants to our clients. When I interview, I want to see written materials the candidates have done. I look for their written skills. How clearly can they explain things? How is their spelling? Do they have computer literacy?"

CABLE CARTEL

Using the available cable networks that offer local ad insertion, combined with the ability to "cherry pick" systems throughout the Southeast, Cable Cartel, based in Fort Walton Beach, Florida, provides targeted customers for advertisers. "We give advertisers the same coverage area as broadcast television at only a fraction of the price," says Bill Sadler, vice-president.

Cable Cartel sells ads on nine different channels—but not necessarily in the same way.

> It's expensive to buy equipment that lets you select the time your commercials run. We have time-selection on four channels. The others are ROS—run-of-station. For those, rates are cheaper.
>
> Rates vary, too, by the channel and time you select. The very same commercial sold as ROS on CNN may cost an advertiser $30, but on ESPN, during a Saints football game on Sunday night, $250. There's a strong audience that follows the Saints.

Each advertising contract is negotiable, depending on the schedule, he says. Advertisers that run with high frequency can get a lower rate.

"Through cable's capability for narrowcasting, we can hit upscale areas—a technique called 'cherrypicking.' An expensive beach resort, which wants to reach viewers who travel and who can afford to stay there, won't run an ad on a channel that's seen primarily in a low-income urban area. Instead, they'll look for a channel whose viewers fit the desired demographics."

The "less-hassle" feature of interconnects make them successful, Sadler says. "Advertisers get a cheaper rate. They don't have to negotiate with 15 people, or send 15 checks out. They send us one tape with the commercial. We dupe the tape—reproduce it—and have it sent to the appropriate cable channels."

Cable Cartel also is responsible for traffic—tracking the schedule on which an order is placed as well as the time the commercial will run. Cable Cartel fills out a traffic report and sends it to the ad agency's traffic department, letting them know the start date, the end date, and the channels selected for each commercial. If the commercial is to run at pre-selected times, the report also notes the programming—the commercial is to run during ESPN's college football games, for instance.

Both part-time and full-time staff are used at Cable Cartel. Part-time work is strictly commission at 15 percent, Sadler says.

Full-time staff, generally hired with advertising experience (though not necessarily in cable), receive a base salary plus commission. Neither part-time nor full-time personnel receive benefits; they're considered outside contractors. Although travel is involved, they use their own cars. "They go on the road, finding companies with products that might use cable advertising in their marketing strategy," Sadler explains. "They suggest those companies use cable in the regional area."

Jobs in traffic require dealing with lots of paperwork. "We keep track of all the current rate sheets, the personnel changes, the accounting, and the billing at the end of each month," Sadler says. Cable Cartel uses a secretary/bookkeeper. "Computer and word-processing skills are a must," Sadler says. "And you need the flexibility to adapt to new technology."

CABLE MEDIA CORPORATION

"We make cable advertising easy," says Barrett J. Harrison, president of Cable Media Corporation, a firm selling commercial cable television time for the Detroit Cable Interconnect, cable systems in 400 cities coast to coast, and the Pro Am Sports System—a regional sports network. "There's no more unreturned phone calls, no more time wasted trying to contact the right people, and no more staying late just to wrap up a hard-to-place cable order."

Harrison formed Cable Media in 1982, after an 18-year career of sales and marketing experience at the Ford Motor Company in various domestic and international assignments. In addition to positions at the Ford Division and Ford Motor Company World Headquarters in Dearborn, Michigan; and with the Ford Division field sales organization in Los Angeles, Salt Lake City, and Philadelphia; Harrison served as marketing manager of Ford's South African operations with Ford International.

Cable Media provides advertisers with easy access to ad-supported networks such as ESPN, CNN, TNT, MTV, and Lifetime, as well as to regional cable sports networks nationwide. In 1988, Cable Media became the first cable rep firm to go international by establishing a sales office in Toronto. The company offers advertisers over 825,000 homes on the Detroit Cable Interconnect, in excess of 750,000 on Pro Am Sports, and nearly 40 million homes nationally.

Harrison, whose company received the prestigious Carl Weinstein Sales Achievement Award from the Cable-television Advertising Bureau in 1992, sees plenty of opportunity for young people in cable advertising.

> The industry is exploding. Young people looking for entry-level positions as sales assistants need the right attitude and the work ethic. They should want to learn the business. A marketing background helps, but you don't have to have it. We've had good luck hiring people with no background in the cable industry.

Harrison looks for three or four strong characteristics in applicants.

> How do they talk? What was their background?
> When we're hiring account executives (who do the actual selling), we look for men and women who can handle an interview well, speak well, and have a nice appearance. We pay them $2,000 a month as guarantee, but they basically sell on commission. We have a package set up. They get 10 percent of the first $30,000 worth of advertising they sell; then 5 percent of everything after that.

Harrison likes to see sales experience, preferably in magazine or in radio. However, the best salesperson he has, he says, had been selling copy machines previously. "We just had to teach her our product," he recalls.

Account executives are assigned a list of from 5 to 15 advertising agencies. "It's their job to present our product," Harrison says. "They establish a relationship with the agencies—create an environment that can maintain a long-term sales relationship.

"Know the industry," he says. "Do your homework. Read the trade journals. Then, when you come in for an interview, you can speak intelligently about cable."

One thing that turns him off on applicants, Harrison says, is a resume that has typographical errors.

> I can't believe the trash I see. If your resume has a misspelled word, when the interview is over, I throw the resume in the wastebasket. If you don't take time or thought in preparing a perfect resume, you won't do a good job for me. Your resume is the single most important document I look at when I make hiring decisions.

His advice: keep your resume to a single page, and do it in chronological form, with your most recent experience at the top.

After the interview, he advises, be persistent without being a pest.

> It's a fine line, and a hard one to follow. You want to send a thank-you letter that's well-written. Be sure you have the correct spelling of the name of the person you're writing to, as well as the title. If you don't hear within two weeks, then make one phone call. That's all.

TECHNOLOGY

The physical process of getting the ads on cable television may be changing rapidly, as new technology is developed and tested. *Cable Avails,* the cable TV advertising monthly trade magazine, reported in 1992 that virtually all 1,750 ad-inserting cable systems were using VCRs—3/4-inch tape decks, but many were looking at alternatives. One possibility: laser disc insertion technology. In one test, a Texas cable system planned to take source tapes, feed ad copy from them to a laser disc machine, and create spot play discs. Next, operators at the cable system would insert the discs into a play-only unit. One laser player is needed for each network on which the system inserts advertising. Laser disc systems give better picture quality than VCR tape, but cost substantially more.

Even more desirable—if feasible—is video signal compression. Under that as-yet-unproven technology, commercials would be coded digitally, stored in a computer, retrieved, and aired. The technology is similar to the way in which a graphics file is stored on a computer, to be printed out later. However, typical commercials require too much computer memory for the technique to be practical. Techniques to compress the signals so they can be stored in less computer space are being developed.

You can keep up with developments by reading the trades. Especially recommended: *Cable Avails* (a cable TV advertising monthly); *Cable World,* which includes news of developments in cable advertising; *Multichannel News;* and *Cablevision.* Addresses for all of them are listed in appendix B. Several have "people" news—columns which list brief announcements of promotions and job-changes.

REACHING VIEWERS

Time Warner CityCable Advertising—an interconnect—gives advertisers an economical way to reach prime New York City customers with the full impact of television advertising. The two cable systems serving Manhattan and Roosevelt Island (Manhattan and Paragon) deliver over one million viewers in almost 400,000 households, 40,000 hotel rooms, 3,500 corporate offices, and 900 restaurants and bars. The Queens and Brooklyn systems reach well over 700,000 viewers in almost 350,000 households and 368 restaurants and bars, plus more corporate addresses in the New York City outer boroughs. Advertising campaigns can run in Manhattan only, Queens and Brooklyn alone, or throughout the entire Time Warner CityCable system.

"Geographical targeting means advertisers don't pay to reach viewers they don't need to reach," says Larry J. Fischer, president, Time Warner CityCable Advertising. "Local cable advertising gives them the flexibility to reach select segments of viewers in specific areas. In addition, they can increase the power of their advertising budgets by using value-added direct mail and print ad services. For instance, Time Warner's New York 1 News offers advertisers cross-channel promotion, pages in the Metro editions of Time, Inc. publications, daily newspapers, radio, outdoor advertising (bus-sides and backs), bill inserts and direct mail to subscriber households."

CityCable offers advertisers 22 networks on which to advertise. Fischer describes the service:

> Our media packages are tailored to advertisers' needs. I've had great success hiring former radio salespeople because they're used to selling packages. Radio has morning and determine drive times, daytime, and nighttimes. In radio, advertisers can't just buy what they want—they have to buy what stations want to sell.
>
> A cable advertiser may want to buy the Arts & Entertainment Network. But we've got to sell the agency four additional networks. Persons who have successful radio experience know how to do that.

Overall, Fischer prefers hiring advertising veterans—not just those with cable experience—over newcomers. At entry-level positions, he requires a minimum of a year or two's experience in a related field, such as an ad agency or a media outlet. "I'm looking for persons who want to dedicate themselves to an industry they believe has great potential," he says.

"When I'm interviewing, I always ask applicants: 'Do you have cable TV?' 'Do you *like* to watch television?'"

"If they answer, 'Yes, I watch cable, but not all that much,' it's a turn-off. I want them to tell me their favorite networks."

Fischer says he doesn't need TV junkies. But successful applicants must understand the cable universe, he advises.

> It's not good enough to know five bullet points about why a particular network like BET is better. You have to be able to know why it's important to advertisers to buy any single one of the networks we sell on.
>
> If I asked my sales associates what's on at 8 P.M. on Lifetime, and they didn't know, I'd be appalled. They should know they can sell 17 NFL football games on two networks. They have to know

where SportsChannel is. They have to know on what channels the Mets games are seen.

Maturity is an asset in hiring, Fischer contends. He likes to hire men and women in their 30s or older, rather than new graduates, and has several staff members in their 50s.

> Advertisers have an incredible number of media choices today. In order to sell successfully, account executives must understand cable's relative value among those choices. That's knowledge that can't be book-learned. Yes, you can show someone a rate card and talk in terms of rates, but unless you've been living advertising, selling advertising, selling against various advertising media, you won't do as well.
>
> I need people to say, "You may be better off not buying a $40,000 ad page in the New York Times, but taking a $30,000 ad. With the $10,000 you're saving, I can do things for you. Wouldn't you rather have 50 $200 spots on CNN or ESPN?"
>
> I need successful people who really like and understand the industry. I want people with a passion for cable.

CHAPTER 8

PROGRAMMING

One of the least understood areas of cable is programming. Many communications majors who concentrate on production hope to land a job in cable where they are responsible for creating programs. But that's not easy.

As we described earlier, cable is basically divided into several parts: cable operators, cable suppliers, and suppliers of cable support services. The *operators* seek out the franchises in the communities or buy them from other cable systems, build new plant, and operate the cable systems. In most cases, they also determine what programming will be shown on the systems. In the case of MSOs, programming decisions are usually made for individual systems at the MSO headquarters.

The *suppliers* supply the programs via satellite. Sometimes they themselves create what is seen over the air. At other times, they buy classic (old) movies or reruns. They can purchase original programming from outside companies, participate in joint ventures, or commission programming to meet special needs.

Cable support services create and/or distribute supplementary materials. Sometimes these may be marketing tools: bill inserts, promoting pay-per-view events or the Olympic triple-cast programming. Sometimes they are program guides, such as those offered by the Arts & Entertainment Channel, The Disney Channel, The Discovery Channel, or American Movie Classics. Ads on the "matching" cable channels may promote mail-order subscriptions to the guides. Some program guides are sent free to subscribers.

Several networks have developed Cable in the Classroom kits which are free or low-cost, available to educators. PBS offers support materials, including teacher's guides, multimedia kits, and posters. Assignment Discovery offers a newsletter with programming schedules. C-SPAN in the Classroom has teachers' guides, newsletters, and lesson plans. Bravo offers program descriptions, reading lists, and discussion topics.

Almost certainly, you will find it difficult to land an entry-level job in creating cable programming. However, if you have writing and editing skills, you may be able to find work in creating support materials.

STEP ONE

If you'd like to work in cable programming, the first step is to start at the local level. Virtually all cable systems have some local origination programming. Almost all of them would like to have more.

Local origination is a program that the cable operator produces. It may be the high school basketball game or band concert. It may be a studio set and "talking heads" having a political discussion, in which mayors or local newspaper columnists discuss community election issues and results. At Continental Cablevision of Northern Cook County, a five-suburb system outside Chicago, programs have featured local storytellers, "Meet the Artist" (which showcases area artists describing their work and demonstrating techniques); "Campus Camera" (from the local community college); and "Hometown Edition," a half-hour show in magazine format, covering the communities the system services.

There's *access programming*—free public time. Normally, this is done by community groups and organizations, or governmental bodies, like schools. Sometimes a state representative will speak, or the area's representative in the U.S. Congress will discuss legislative issues that have local impact.

Most franchise agreements, especially those in smaller communities, require cable operators to set aside channels for local origination and public access.

Some cable systems are required by their franchise terms to offer free or low-cost workshops in cable production. Residents who sign up for classes are trained in scriptwriting, camera work, videotape editing, and other production tasks. Once residents have been "checked out" on the equipment, they may borrow it and create programs that are then shown on the air.

Programming opportunities do exist at the local level, but producers here, for the most part, are volunteers—*not* making television shows the way the supplier networks do. Instead, they're working with the community, using cable as the communications link, as a way of reflecting the community in its own people's eyes.

If you're happy with that kind of orientation, if you don't look on it as a springboard to fame and fortune, and a network contract, then there are opportunities. In fact, technology is providing even more chances for local programming. As older systems upgrade, as cable systems (even in smaller communities) begin to carry increasing numbers of channels, *someone* must offer programming to fill them. That "someone" could be you.

A SMALL CABLE SYSTEM

At cable stations in smaller communities, people often have more than one job function. Bishop, California (in the Owens Valley on the eastern side of the Sierras) is served by WESTSTAR Cable Television. "Ted produces commercials and engineers the news," says Susan Hook, the station's general manager. "He's also weatherman on the news.

"Lynn oversees our live programming. She schedules guests, makes sure the hosts of the various shows are communicating with each other on guest lists and schedules, and co-anchors the news. She has her own beat. In one day, she may go to the sheriff's department, the California Highway Patrol office, the Department of Motor Vehicles office to see what's going on."

In the bigger cable systems, Hook says, personnel are far more specialized. "All someone may do is interview," she explains. "But here, we cross-train. We double on jobs."

At Bishop, 23-year-old Matt McGee is news director—a job he calls "similar" to that in a broadcast station. McGee graduated from Pepperdine University as a broadcasting major, interned six months at a Los Angeles station in the sports department, and worked six months for the *Los Angeles Daily News* as an editorial assistant before being offered the job for the Bishop cable station.

Here's how he describes a recent day:

> Putting the news together takes up nearly all my time. Our 30-minute show runs five nights a week. I'm producer, writer, reporter, and anchor. We have a three-person department: two journalists and one weatherman. The weatherman doubles as our technical director—pushing all the buttons and rolling the videotape when we have something spectacular like a fire. We videotape his weather segment in advance, so he can sit in the booth while the show airs and run the technical equipment.

McGee's "typical" day involves much phone contact with local sources because the station is too small to have an AP wire service feed. He's on the mailing list of various state and county agencies—receiving press releases and background information, which he turns into stories with a local angle.

He'll sit at the computer and write stories, he says. He'll call a radio news service in Washington, D.C. and get "sound bites" from California senators. He'll decide which stories are worthy of video, and which aren't.

If a news story breaks locally, McGee will take a camera and go out to shoot footage. "None of us could survive on one job description alone," he points out. "If I'm tied up, even our installers can go out and shoot. We all cover for each other."

By 12:30 P.M., news gathering has finished for the day. McGee typically spends the next two hours editing the video together for that evening's broadcast—determining the importance of each story, assembling the entire newscast. By 3:30 P.M., he and his two colleagues have their makeup on and are in front of the cameras, taping the newscast.

Days are long, he says, and he works most weekends. "All the press releases have to be looked at and filed," he explains. "If I didn't go into the office, I'd be too far behind, and couldn't catch up."

McGee says only four persons from his college group of 40 at Pepperdine received job offers when they graduated. He credits his experience with the student radio and television stations with making the difference.

> Hands-on experience is vital! All the *A*'s on a transcript aren't going to get you a job if you don't know how to turn a camera on and take pictures. You'd better know how to shoot your own interviews and use an editing machine. Working at the student station also gives you a chance to make a resume tape while you're still in school.

Cable programming wasn't an area McGee thought he'd be in. "Everything I learned in college told me my first job would be in a tiny town, where I'd make $9,000–$12,000 a year to start," he recalls. "I was a broadcast major. We rarely mentioned cable. But because this is a resort community, because salaries are commensurate with living costs in the area, I started at $1,500 a month."

His advice: "Show prospective employers how you can make their jobs easier. Be curious. Be aggressive. Sell yourself. You need to present a confident attitude, and make people know why they should hire you."

CREATING PROGRAMS

Would-be programmers who think their communications degree will open up programming jobs right after college will probably be disappointed. Most cable executives say it just doesn't happen. Here's a warning from one executive:

> If you happen to be standing at the right place at the right time, having the right qualifications that someone needs—maybe. I

know young people to whom that has happened. But pretty generally, their father is the chief executive officer of a company, or a major entertainment lawyer in Los Angeles who calls up five people and says, "Interview my kid."

For the most part, however, cable operators don't create programs. Instead, production in cable generally comes from *satellite program services*. You'll find an up-to-date listing of these services, along with a description of each, in *Producers' Sourcebook: A Guide to Cable TV Program Buyers*. Write to the National Academy of Cable Programming, 1724 Massachusetts Avenue, N.W., Washington, D.C. 20036 for purchase information.

This time-saving handbook, updated annually, contains information on 61 national cable programming networks and 27 regional services. Each network profile provides current data on programming content and formats; guidelines for program acquisition; technical requirements for submissions; and contact names, addresses, phone and fax numbers. The *Sourcebook* indicates who is buying what, and how to reach those buyers. In 1991, for instance, Court TV wasn't interested in independent productions. Comedy Central was buying approximately 1,300 hours of original programming per year. And The Learning Channel was buying 85–90 percent of its programming. Sunshine Network, an Orlando-based regional network with 95 percent of its programming in sports, also co-produced college football, basketball, championship events, professional tennis, boxing, and single sporting events.

NETWORK PROGRAMMING

Networks also profile themselves in the media kits they prepare. Black Entertainment Television (BET) says it's committed to excellence in urban contemporary programming. Launched in 1980, this network, seen in 33.4 million households over 2,500 cable systems in the United States, Puerto Rico, and the U.S. Virgin Islands, was the first network to showcase quality black programming 24 hours a day.

"Our programming is an exciting mix of today's hottest music videos, jazz, gospel, action-packed sports, news, public affairs, star-studded specials, and much more," says Jeff Lee, vice-president, network operations. "In addition, we own a majority interest in *Emerge,* a monthly newsmagazine for middle-income black people, and have developed *YSB (Young Sisters and Brothers),* a magazine for teens."

About 65 percent of BET's programming is sent by the record industry as music videos, Lee says. "We repackage it into different groupings: 'Rap City;' 'Video Soul' (the hits as they appear on the rhythm and blues charts); 'Midnight Love,' with romantic music; or 'Video Vibrations,' which groups popular music from different genres."

Some news and public programs like "Our Voices" are produced in-house at BET's Washington, D.C. production facilities, Lee says.

> We built the studios in 1989 as a way of controlling costs and product distribution. Our staff includes everything from set builders to executive producers—from production assistants to associate producers and producers. We have researchers, writers, set builders, and technical jobs from a director of engineering down to a TelePrompTer operator.

Lee says production jobs include wardrobe, makeup, hairstyle, and set construction—while programming jobs include writers, directors, and producers.

BET covers sports stories as part of the African-American life-style, Lee points out.

> We try to showcase black college football. But our audiences see more footage of the bands than other networks show. We show the experience of going to a game at a black college as part fashion show, part concert, and part football.
>
> Black basketball shows include African-American journalists talking about issues and topics of the day. Sports profiles are shows featuring information about retired athletes.

BET's distinctive programming is designed to reflect the needs, interests, and diverse life-styles of black America. But Lee says the network doesn't count color or gender when it hires.

> We're looking for self-motivated people who have a drive to achieve—not people who have to be told what the next step is. We want self-starters. We want people who get gratification from doing a good job.
>
> They analyze jobs in terms of where they can start and what they have to do. They do their work quietly; they don't have to parade their success.
>
> When I interview people, I tell them: "If your goal in life is to do exactly what this job you're applying for is, I don't want to hire you. If your goal is to expand this job, expand your knowledge to go on to do something else, you'll be successful." These are the people we like to have around. These are the people who are good for our company—and good for cable.

CHAPTER 9

SCHOOLS AND TRAINING

If you want a job in cable, choosing a school and a course of study is not necessarily easy. Because opportunities in cable are growing rapidly, most universities have not set up comprehensive cable programs.

Widely known schools like the University of California, Los Angeles (UCLA), the University of Southern California (USC) and Boston University have only one or two cable-related courses even though they offer majors in film, video, or television; in communications; or in broadcast management. Today's cable executives, however, have often reached their positions without a formal background in cable television studies.

WHAT SHOULD YOU STUDY?

"Too many students are too narrowly focused on communications career sectors," says Frances Forde Plude, associate professor at Syracuse University's S.I. Newhouse School of Public Communications. "Students need a broader view of employment opportunities." Plude suggests that telephone companies will play a major role in the cable sector's future. "Other new technology aspects like direct broadcasting satellite (DBS) and high definition television (HDTV) will also interface and overlap with cable market sector activities," she says.

Because cable's role is changing so quickly, there is no "right" course of study that can automatically assure you a job in cable when you graduate.

Perhaps the first thing you should do is to consider carefully your talents and skills, your own likes and dislikes. Are you strong in math and electronics? Do you enjoy working with electrical equipment? If so, your prospects may be bright.

Or perhaps you are interested in business and management. Cable systems need managers at the local level who can supervise customer representatives and handle office problems. Courses in accounting, management, marketing, and personnel can help you land such jobs.

Advertising is a growing field for cable systems. General business courses which include some exposure to media advertising sales and methods may help you here. You'll want to learn about demographics, about market research, about ratings, and how audiences are measured in cable.

Production is another field of cable that seems glamorous and exciting. But film and video production majors may not find it easy to get paying cable jobs unless they have other skills or strong connections. Most cable systems, especially at the local level, do not produce original programming for most of the shows they broadcast. Instead, they buy such services from outside agencies or networks.

Major networks often commission programming or work closely with independent production companies. For instance, in 1991 Turner Network Television spent approximately $75 million to develop and produce 24 original movies/miniseries. HBO co-produces virtually all types of original programming. The Arts & Entertainment Network (A&E) generally contracts approximately one-third of its budget to independent producers. The network purchases and co-produces original programming in documentaries, drama, and performing arts.

You'll find a list of cable program suppliers in *Producers' Sourcebook*, available for purchase from the National Academy of Cable Programming, 1724 Massachusetts Avenue, N.W., Washington, D.C. 20036.

Persons with film or video production backgrounds can sometimes be hired to coordinate local access programming in individual municipalities. But the chances of jumping from such a job to a programming position in New York or Los Angeles are almost nonexistent.

HOW TO FIND OUT WHAT'S AVAILABLE

Because there's no direct, clear-cut path to cable success, you must be unusually aggressive in learning about cable training and education.

A good source of information on schools is *The American Film Institute Guide to College Courses in Film and Television* (8th ed.) This booklet lists several hundred courses in motion picture and television broadcasting offered at schools throughout the United States, along with addresses and names of department chairs.

You'll need to write for a variety of school catalogs and study them carefully. Contact school officials before you enroll. Ask hard questions about intern opportunities, placement rates, program graduates who are actually working in cable.

"Careers in Cable Television," a booklet available for purchase from the National Cable Television Association, 1724 Massachusetts Avenue, N.W., Washington, D.C. 20036, describes a number of cable jobs in detail. Included are jobs in management, technical, administration, marketing and advertising, programming, sales, and legal areas. As you read the description of responsibilities for each position, imagine what knowledge you must have and what skills you must demonstrate to qualify for such a job.

Read the trade publications to learn cable trends and issues. You'll find a list of periodicals in appendix B. Subscription rates to most of them are reasonable. Your library may be able to get back copies for you through interlibrary loan services. Or, your local cable television system may be a subscriber. While a cable system probably won't lend out current issues, a manager may allow you to sit in the reception room and read. Ask if you can have any discarded copies.

Trade publications will also give you lists of conferences or conventions in your area. Frequently, you can attend these meetings for a nominal student fee. Cable executives suggest you go there. "Speak to programmers and hardware suppliers," they say. "Talk with system operators. You'll have a greater awareness of cable, and be better able to make decisions."

CUTTING THE COST

Although recent cuts in financial aid have made it less likely you'll qualify for grants or loans, begin early to check all possible sources of money. Talk to your school counselor about scholarships. Fill out financial aid forms early. State agencies award help on a first-come, first-served basis until the money runs out. Check with your local area financial institutions; one bank may not be making student loans, but another may be willing to help.

Some colleges with strong cable programs are state schools, such as Michigan State University or Indiana University. Others are community colleges. If you are a resident of the state, or of the community college district, tuition may be low.

Minnesota, Wisconsin, and North Dakota have tuition reciprocity agreements. Residents of each can enroll in programs offered at state-supported colleges and universities within the other two states, and are required to pay only the "resident" tuition rate. If you live in one of those states, check with your school counselor, as well as with the admissions and financial aid offices of colleges you're considering.

You'll have to check residency requirements carefully, school by school. But you may find that moving to another state and working for a year, so that you become a legal resident, may financially be worth the trade-off in time before you start your education. In addition, your experience may make you more mature, and better prepared to take advantage of your educational opportunities.

If you enjoy the technical side of cable television, if electronics and related fields interest you, you'll want to explore education and training for these kinds of jobs.

Technical training programs prepare students for entry-level positions as installers or technicians. Some technical training schools are trade schools; some are skill centers, and some are community colleges. Most have financial aid available and an "open door" admissions policy. Graduates of electrical engineering courses at major universities often find their training valuable for cable positions.

Opportunities for learning about television can be found in almost every state. Not all schools which teach television or television production courses, however, deal with cable.

If learning about cable television is your goal, you will need to go through the *Guide* carefully, school by school, to find out which colleges and universities offer cable-related courses, and how many. Often an institution with a television major will list only two or three courses in cable. If a school interests you, write to the television department chairperson for up-to-date information. Because of cable's increasing impact, colleges are adding new courses which may not be shown in the *Guide*.

Since opportunities in cable production are extremely limited, you may want to consider combining business administration courses (sales, marketing, advertising) with cable television courses. Ask the admissions officer at schools you're considering whether you can do this.

SCHOOLS

Here are some schools that offer cable-related courses:

Arkansas State University, Jonesboro, AR: This four-year state university offers a number of broadcast news, production and practicum courses that give students a chance for direct cable experience. The University uses a cable channel assigned to it by United Artists Cable in Jonesboro, as well as a campus cable channel. Advanced broadcast news students serve as reporters, producers, and on-camera anchorpersons for a 30-minute afternoon news-weather-and-sports block.

Students in Advanced Television Practices provide the in-studio production crew for the newscast and produce additional special television programs for the cable channel.

A sales practicum course assigns students to local radio, television, and cable sales departments, as well as to local advertising agencies, under the supervision of sales or advertising personnel. United Artists Cable cooperates with the department by accepting students for summer internships and the sales practicum course.

At ASU, five instructors are teaching cable-related courses. One popular course is Cable Systems Managements and Programming, which includes the study of cable management responsibilities and cable regulation. It is designed for students who are preparing for management and executive positions in single or multisystem cable television operations. A number of the university's alumni are working in cable-related fields or with cable systems.

For information, contact the Chairperson, Department of Radio-Television, P.O. Drawer 2160, State University, AR 72467.

Columbia College, Chicago, IL: This commuter college emphasizes education in the arts and media, with degrees at both undergraduate and graduate levels. The faculty for arts and media courses are working professionals. Building on fundamental skills learned at Columbia College, alumni are working in production crafts (producing, directing, and writing); in administrative positions (programming, sales, promotion, and research); and in talent areas (commercial announcing, news anchoring, reporting, and performing).

At the undergraduate level, there are BA's and departments in film/video, television, journalism (including broadcast journalism), management (with special attention to media and arts management), and marketing communications. Columbia also offers a Master of Fine Arts in film and video.

Students may also design their own programs by combining courses from related departments at both undergraduate and graduate levels. Courses include such topics as film and video techniques, optical printing, animation, film and video editing, and cable production.

Study courses typically include workshops, independent projects, and co-op projects. A variety of internships are available; past internships have included positions in cable. Television students produce three broadcast format programs distributed throughout the college on Access Cable, and several other systems. Studio facilities are extensive. Financial aid is available.

University of Denver, Denver, CO: The University of Denver's School of Communication offers an undergraduate degree in communication with a telecommunication emphasis (both cable and common carrier). Course offerings include: Telecommunication Management, Telecommunication Economics, Telecommunication Law, Telecommunication Systems & Engineering, and a Seminar in Broadband Program Delivery Systems.

At the graduate level, the Department of Mass Communications and Journalism Studies offers information about cable in a number of courses, including: Finance and Management of the Mass Media and Issues in Mass Communications. It's possible to earn an M.A. in Mass Communications, M.S. in Public Relations, and M.S. in Advertising Management.

The Department of Mass Communications and Journalism Studies and the School of Communication have a strong internship program which allows students to take advantage of Denver's location to pursue a variety of internships in the cable/telecommunications area.

Indiana University, Bloomington, IN: A strong, nationally recognized degree program in telecommunications is available through the College of Arts and Sciences. The program has proved so popular that special admissions requirements are being developed. Check with the Telecommunications Department for up-to-date information.

For university admission, Indiana residents must rank in the upper half of their high school class, and have a total SAT score above 850, or comparable ACT score. Out-of-state residents should rank in the top third of their high school class. They should have attained a total SAT score of 950, or a comparable ACT score. Freshmen generally spend their first year in the University Division of the College of Arts and Sciences. Qualified students may test out of many of the required courses for graduation, including writing.

Telecommunications majors receive a strong liberal arts education plus in-depth instruction in all aspects of the electronic media, including broadcast, cable, and telephony. Majors may specialize in one of the following areas of coursework: telecommunications industry and management studies; electronic media production; information technologies; and telecommunications, society, and culture. Alternatively, students may devise their own plan of coursework from among all four topic areas. Depending on a student's career focus, study in any one of the four topic areas would support a career in the cable communications industries.

The Department offers several courses specifically geared to the study of cable television, plus many other courses that relate to various aspects of the cable industries. These courses include Cable/Broadband Communication, Broadcast and Cable Promotion, Programming Strategies, Electronic Media Audience Analysis, Television Production, Processes and Effects of Mass Communication, Television Aesthetics and Criticism, Content Regulation of Telecommunications, and Structural Regulation of Telecommunications. Additionally, the Department administers a strong internship program in which telecommunications majors earn academic credit by working in the industry for a semester or a summer.

Graduate study leading to both the Master's and Ph.D. degrees is available in the Department. Of particular interest to those seeking careers in cable management is the Master of Science degree, which concentrates on electronic media management.

For further information about graduate programs in telecommunications, contact the Graduate Coordinator.

Ithaca College, Ithaca, NY: This four-year comprehensive college offers three degrees relevant to careers in cable television. The first, in television/radio, offers students opportunities in audio and video production and media management. Other areas of study include advertising/public relations and scriptwriting. The second degree, in telecommunications management, includes a strong emphasis on both communication and business courses with a strong liberal arts element. The third degree, in journalism, allows students to emphasize broadcasting and cable-oriented news.

Students in all three degree programs have outlets for their creative work on an access cable channel shared by educational institutions in Ithaca. During the average week, Ithaca College Television originates 22 programs, with some being available through satellite delivery systems. Many of these programs have won national competitive awards.

Mercer County Community College, Trenton, NJ: This college offers a 2-year Associate in Applied Science (AAS) degree in television production or radio. Classes feature intensive hands-on training for on-air, management, and production responsibilities. The Television option prepares students for beginning jobs in TV and cable TV production—from camera operator and lighting technician to video editor, writer, and TV director. Students gain additional production experience in the Mercer College Cable Network and the Mercer College Public Radio Network. Qualified students have the opportunity to intern at cooperating cable TV systems, TV and radio stations, or other production facilities in the New York City, Philadelphia, and New Jersey areas.

Middlesex Community College, Middletown, CT: This college offers a two-year associate degree or a one-year certificate program in Broadcast Communications. The program emphasizes training in television production, but also covers radio, cable television, film, corporate and educational video, and general communication studies.

Middlesex has broadcast production facilities. They include a fully equipped television studio, audio and editing suites, computer graphics stations, and a student radio station.

Besides general introductory production courses, the program offers specialized courses in advanced A/B roll editing, audio production, computer graphics for video, broadcast journalism, performance, and script-writing, as well as an internship program.

About two-thirds of program graduates obtain jobs at broadcast TV stations, radio stations, cable companies, corporate video departments, or private production companies. Most other students transfer to four-year degree programs.

For further information, contact Broadcast Communications Department, Middlesex Community College, 100 Training Hill Road, Middletown, Connecticut 06457.

Michigan State University, East Lansing, MI: An intensive course sequence in telecommunications prepares students for management positions in telecommunication industries, including broadcast stations, cable systems, and common carrier communications. Students may choose courses in the area of production and programming, sales and management, cable communication, telephony and information services, international telecommunication, and telecommunications policy. Students can get experience at the university's public television and radio stations, and telephony laboratory. Chances for internships are available.

At the graduate level, the Department of Telecommunication offers work leading to the Master of Arts degree, and participates in the doctoral program in mass media.

Graduate students study message creation process and effects, the systems and process of international telecommunication; the economics of telecommunication; the organizational structure and management of telecommunication systems; criticism and review of media content; informational technology systems (telephony); development and analysis of public and private telecommunication policies; and the designation of appropriate research strategies for solving problems and clarifying issues.

New York University, New York, NY: This private university offers professional training in all aspects of the performing and media arts through the Tisch School of the Arts. The school's interactive telecommunications program focuses on the development of emerging multimedia and applications of interactive telecommunications systems.

Undergraduate programs in television, radio, and photography are offered. The Department of Cinema Studies and the dramatic writing program are open to both undergraduate and graduate students. A combination MFA-MFA in Film and Television is offered to exceptionally qualified students.

Several undergraduate courses in cable are taught. Both historical/theoretical and practical themes are emphasized. For instance, Strategies and Programming for Broadcast, Cable TV, and the New Technologies discusses bases for decision making, program positioning, competitive- and counter-programming, and mass and specialized audiences. Another course deals with advertising, audiences, and marketing for broadcasting and cable.

At the graduate level, the university offers a two-year program leading to the degree of Master of Professional Studies in interactive telecommunications—the first of its kind to be offered. Competitive admissions take into account diverse backgrounds of students. Studies in cable television, videotex, satellite communications, video teleconferencing systems, and similar topics are available.

St. John Fisher College, Rochester, NY: This independent, coeducational liberal arts college has a communications/journalism major with some cable television courses. Fiscal management, regulatory history, and programming are stressed. Internships are available with American Cablevision of Rochester, New York.

Southern Illinois University at Carbondale, Carbondale, IL: The radio-television department teaches several cable courses, including a survey of cable communications and a workshop in creating and producing cable programming. One course is an internship, which can be taken with a cable, network, or public station.

Syracuse University, S.I. Newhouse School of Public Communications, Syracuse, NY: Students here learn about the broader view of the cable market, including the integration of communication fields and technologies. For example, the course "Cable and Broadband Communications" touches on telephony, computers, integrated services digital network (ISDN), and fiber optics. The course also has an international focus because the school feels that the European Community and Japan telecom markets will have a great impact on today's students' job search. For information, contact the S.I. Newhouse School of Public Communications, 215 University Place, Syracuse, New York 13244–2100.

University of California, Los Angeles, CA: Within the School of Theater, Film and Television, courses in television and video production are offered, along with screenwriting workshops. Undergraduate students enter at the junior level and must submit a written portfolio for evaluation. At the graduate level, several programs are offered. All lead to a Master of Fine Arts degree: A three-year program for independent filmmakers and video artists, a separate two-year program in screenwriting, and another two-year program for film and television producers that emphasizes business skills. In addition, UCLA offers numerous short courses and continuing education workshops.

Wadena Area Vocational Technical Institute, Wadena, MN: One of only two colleges in the state of Minnesota offering Cable Television Technology, the program at WTC has been offering technical training in this field since 1974. WTC offers two training paths in cable television: an 11-month Cable Television Instal' program and an 18-month Cable Television Technician program.

Students must be at least 16 years of age to enroll. A tuition reciprocity agreement with Wisconsin and North Dakota qualifies residents of either state for the Minnesota resident tuition rate. Financial aid is available to qualified students through various means, including short-term loans, grants, and work-study jobs. A fully licensed campus day care facility is also available for students with children.

Training in the Cable Television Technology field combines basic electronic knowledge and theory with hands-on skill practice. The curriculum includes basic electronics, semiconductors, radio frequency (RF) transmitter and receivers, digital, CATV line equipment, headend equipment, system design, RF test equipment, microwave, and fiber-optics training. Construction training takes place outside in a training field with pole line construction and climbing, aerial and underground cable placement, machine operation, and installation of equipment housing.

Wisconsin Indianhead Technical College, Rice Lake, WI: WITC provides a four-semester vocational diploma program, Cable Television System Service. Students construct the outside plant, aerial and buried. They run cable, installing amplifiers and cable taps as needed to activate the system and provide service to customer premises.

The program is supported by AC/DC instruction, Solid State Electronics, and Digital Logic, so that further instruction in system alignment, balancing, sweeping, performance standards, and operating parameter measurements are fully understood.

The program is 1,980 hours in length, and students earn 59 college credits during the two school years. Internships may be earned during the summer between school years. Graduates begin as installer/technician, and have training which will support the goal of achieving higher technical levels. Contact Trade and Industry Supervisor, Wisconsin Indianhead Technical College, for information about the program.

SEMINARS

For students in the Los Angeles, New York, and Chicago metropolitan areas, short workshop courses are frequently offered in screenwriting and related topics. Many deal with video and film production. You can find information on current workshops by reading the classified ads in such trades as *The Hollywood Reporter. Cable World* runs a calendar of events that lists conferences, technical seminars, meetings, and training programs. The list gives sponsoring organizations, contact names, and phone numbers.

Over 70 seminars and workshops annually are presented by the American Film Institute's public service programs. They're held at the AFI campus in Los Angeles, and in cities throughout the United States. These programs are taught by top film and television professionals, and provide the film and video community, the independent artist, and the general public with a forum for exploring current artistic and technical issues. Summer workshops from AFI's Education Services are meant for industry professionals and the general public.

For details, contact The American Film Institute, Public Service Programs, P.O. Box 27999, 2021 North Western Avenue, Los Angeles, CA 90027.

Media analyst Paul Kagan presents a number of seminars on cable-related topics. A recent meeting on "The Future of Pay-Per-View" discussed such questions as "Will the Olympics Triplecast usher in a new age for pay-per view? When will major league sports—football, baseball, and basketball—move to PPV? and Is multi-plexing the answer to low PPV buy rates, or is it still a marketing/pricing issue?"

The Kagan Group of companies also offers seminars on audiocassettes that can be ordered by mail. Topics include cable TV overbuilds/telco competition; future of DBS and pay-per-view, and cable TV values.

For information, contact Kagan Seminars, Inc., 126 Clock Tower Place, Carmel, California 93923.

INTERNSHIPS

An internship with a cable company, a cable network, or a related organization may be another way to get experience. Most of the schools listed in this chapter have internship programs.

Several books about internships are available, and updated frequently. You'll want to read *Internships, 199x: 50,000 On-the-Job Training Opportunities for College Students & Adults,* from Peterson's Guides, as well as two volumes in a series on internships edited by Ronald W. Fry: *Internships, Vol. 1: Advertising, Marketing, Public Relations & Sales,* and *Internships, Vol. 5: Radio & Television Broadcasting and Production.* (See "recommended reading: books" in the appendix B.)

Most interns in cable have studied television production courses or communications courses in college. But that's not always the case.

Graduate student Tom Hotaling, 24, who interned at ESPN, was a history major in his undergraduate days at Siena College in Loudonville, New York. He'd always been interested in sports, and played on Siena's soccer team. Wanting to continue that interest for his career, he enrolled in the master's degree program Adelphi College offered in sports management. As part of that program, he needed to complete an internship during his last semester.

"I'd never taken cable television courses," Hotaling recalls. "I hadn't worked in cable. But I do like to watch sports on cable, and wondered if a sports management background could fit with cable opportunities.

"ESPN and several other organizations had written to Adelphi looking for interns. I answered the letter and faxed a resume. By mid-December, I was invited for an interview at the network's Manhattan offices. By mid-January, I'd been offered the internship. I started February 10 and finished May 15."

As an ESPN intern in public relations, Hotaling worked from 9–5 Mondays through Fridays. He was paid, but received no benefits.

Part of his duties included clipping files and distributing them to department heads and executives, so everyone knew what was happening in the exciting, ever-changing world of cable. Consequently, Hotaling scanned trade publications for items related to ESPN or the cable industry, reviewing the business and sports sections, as well as the TV sports columns. His "must read" list included various New York papers, as well as *Variety, Broadcasting, Multichannel News, Advertising Age, Media Week, Electronic Media,* and *Business Week.*

In a large network like ESPN, there is a lot of work in connection with putting out press releases and getting reports out to the media. Hotaling researched information and contributed to a number of press releases for the network. His writing skills improved during the internship, he says.

Like virtually all interns, Hotaling had "go-fer" duties—answering phones, making photocopies, sending and receiving faxes. Those tasks, he feels, gave him the chance for exposure within the company—dealing with various departments and people, as well as gaining valuable experience.

While Hotaling expected his ESPN internship would help him in finding a job in sports management, he feels the biggest benefit from being an intern was the insight he gained. The internship was an eye-opener, he says, about how complicated a big cable network is, how many people are involved, and how interesting behind-the-scenes-work in a cable network office can be.

CHAPTER 10
JOB-HUNTING TIPS

Finding your first job in cable may not be easy. Even though cable is a growing industry, the economic recession has made it more difficult to get entry-level jobs—not just in cable, but in all businesses.

There are a number of things you can do, however, to better your chances, including getting hands-on experience locally. In almost every cable system, requirements of the local franchise specify that there must be a certain amount of local access or local origination programming produced by area residents.

"Get hands-on experience early in your college life," advises Matt McGee, a recent Pepperdine University graduate who majored in broadcasting.

> In fact, if you're interested in cable, don't even consider a school that doesn't have a television or radio station. Start working there—probably on a volunteer basis—as soon as you can, whether or not your coursework requires you to.
>
> When my classmates and I started job hunting, it quickly became apparent that those who'd had four years of experience at the campus station had a higher success rate than those who'd begun working at the station during their third or fourth years in college.
>
> If you want to work in cable, you'd better know how to use an editing machine—how to shoot your own interviews . . .

All the classroom theory, all the *A* grades in the world aren't going to get you a job in cable if you don't know how to turn a video camera on and take pictures.

"PEOPLE SKILLS" ARE VITAL

You can develop "people skills." No matter what job you have in a cable system, you will be dealing with people. Some jobs, of course, are more people-oriented than others.

Customer service representatives, who may handle over a hundred phone calls a day, must be able to help customers with many kinds of problems. "It's like a complaint department in a store," says one television executive. "You've got to be able to solve as many problems as you can over the phone, while still keeping the customer happy with the company."

Because of the tact and attitude a customer service representative *must* project over the telephone, many cable companies prefer to hire persons with customer contact experience for these positions. Such experience can come in many ways. "Someone who's sold in a department store is used to meeting the public," one executive says. "Customer service experience for the telephone or for other utilities is also a good background."

Prior successful sales experience—which shows that you've demonstrated confidence, product knowledge, and the poise and skills needed to persuade customers to buy—can be extremely helpful in getting a cable job, say industry veterans.

Read the trades and keep up with the industry. Over and over again, cable television executives interviewed for this book stressed the importance of knowing the industry before you apply for a job. Because cable is changing so rapidly, they say, keeping up with trade publications may be the only way to know what is happening.

The list of periodicals in appendix B is just a starter. Make a point of looking regularly for cable-related news in business publications, such as *Business Week, Newsweek, Time,* and the *Wall Street Journal.* Check for "show-business" news, as reported in *Advertising Age* and *Ad*

Week—advertising trade journals that include information on television, newspapers, radio, and other media.

The Hollywood Reporter and *Variety* regularly carry material on cable television developments. Every few weeks, *The Hollywood Reporter* lists all films, made-for-TV movies, and television specials in pre-production, production, and post-production—by company, with the name, address, and phone number of the organizations involved. Reading these lists will help you keep up with names of key personnel, as well as keep you current on what's happening.

Cable networks often produce (or contract for) original programming. For instance, about 40 percent of HBO's programming is original. Nickelodeon's co-productions include children's soap operas, narratives, sketch comedies, sitcoms, game shows, animations, and reality-based shows. USA Network buys approximately 30 movies and 6–8 series per year. And in 1992, Turner Network Television (TNT) spent approximately $75 million to develop and produce 24 original movies or miniseries.

The Hollywood Reporter Blu-book Directory, published annually by *The Hollywood Reporter,* contains over 30,000 listings and 300 categories. It includes film, video, broadcast television, and cable companies and personalities, together with "how-to-contact" information. This "one-stop sourcebook" includes the names, phone numbers, and addresses of production and post-production companies, facilities, support services, studio and network executive rosters, celebrity contacts, unions, and guilds.

USING TRADE PUBLICATIONS EFFECTIVELY

Once you've located magazines you'd like to read, write to the publisher of each one. Ask about subscription rates and prices of a single copy. Many of these publications cannot be found on newsstands because they are so specialized; the only way to get a sample copy is directly from the publisher.

After you've seen the differences between magazines, subscribe to several. Read them thoroughly. Cable executives say it's the fastest way to know industry issues and problems. Another way of seeing sample copies is to ask your local cable company if they'll let you borrow back issues.

Although many trade publications are not indexed in *Readers' Guide to Periodical Literature,* you'll find articles of a more general nature are listed. Check under "cable" and under the various subcategories mentioned. A companion index to *Readers' Guide,* the *Business Periodicals Index* is organized in much the same way. Magazines included in this publication are primarily business-oriented, with heavy emphasis on marketing and finance. Articles on the economic aspects of cable, as discussed by the business publications, will be indexed here. *Applied Science and Technology Index,* another companion volume, often lists articles about technical developments related to cable, such as high-definition TV.

Though it's a bit more difficult to use, *The New York Times Index* will give you references to that newspaper's cable coverage. Frequently, libraries have back copies of *The New York Times* on microfilm. Often, these can be used to print copies of articles through a coin-operated device attached to the microfilm reader. The *Wall Street Journal Index* is another good source of information. Once you have publication dates, you can look up issues that carried specific cable-related stories of interest to you.

There's a *Chicago Tribune* subject index, also, but stories covered by the paper are much more regionally oriented, and may not always reflect the national trends. Check your local library to see if major papers from your area have their own indexes, with past issues available on microfilm.

Think local. It's tempting to assume that your communications degree or technical know-how will land you a spot with a national MSO. But your strongest chance for entry-level jobs lies with your local or nearby cable company. There are several reasons for this.

Many franchise agreements spell out the commitment that the cable company must make to hire locally. Some agreements call for hiring large numbers of women and minorities. Others give preference to local residents. A large number of cable systems will offer free training to residents who'd like to learn to produce shows. Once residents qualify, the cable company will often let them check out equipment and shoot locally. In fact, the cable systems will usually put the shows on the air, if they are well-produced.

If you want to know just what the cable company has said about its hiring commitments, visit the office of the municipality that granted the franchise, and read it.

If planning or producing cable programming is your goal, be realistic about your chances. You may well find a job at the local system level, planning public access or local origination programs. To graduate with a communications degree and say, "I'm going to find a programming job in New York or Los Angeles," is optimistic, but probably won't work.

It's unrealistic to expect instant success as a new graduate. It's highly unlikely. If you want to get into New York-based programming, you can make the rounds of all satellite-delivered program services based in New York. If you happen to be standing at the right place at the right time, with the right qualifications that someone needs, you might be lucky.

DO YOUR HOMEWORK

Research companies to which you're applying. Before you interview, learn as much about them as possible. Cable systems frequently have promotional literature which you can ask for and study. Many corporations, and most MSOs, have annual reports which you can request and study. You'll get an idea of how well the cable system is doing. Frequently the report, or press releases, will talk about coming plans or proposed expansion of services.

Your public library will have business publications like Dun & Bradstreet, Standard & Poor's *Register*, *Moody's Investment Service* (for publicly held corporations), and Thomas *Register*.

Don't forget to check *National Business Employment Weekly*, published by the *Wall Street Journal*, and available on almost any newsstand nationwide. Though you're unlikely to find any cable companies advertising job openings, you'll get good, hard-hitting articles on job-hunting techniques: "Attracting the Attention of Executive Recruiters" (an article which listed the top 50 North American search firms) or "A Guide to Diagnosing What's Wrong With Your Resume." As the paper reminds applicants, "In the employer market, a typo isn't a minor tactical error. It's a sign to employers that you're careless and unprofessional."

It takes more than a resume sent to the personnel department of your cable company to break in, or to even get an interview. Find out—in advance—the name of the person you want to see. Write a cover letter and enclose your resume. In the last paragraph of your letter, indicate you'll contact the person within several days. Then do so.

When you write your resume, remember to put yourself in the place of a prospective employer. Why should he or she want to give you an interview? Be sure to include your previous employers, positions, length of employment, responsibilities, and accomplishments. You can also describe your education, and how any courses relate to the available openings.

The purpose of a resume and cover letter is to get you an interview—nothing more. They won't produce a job offer. And you can answer specific questions at the interview. Consequently, most career experts advise that you downplay your volunteer or community involvement in a resume, unless that involvement has given you skills related to the job being filled.

When you contact a cable, or cable-related company, be flexible—not only in your cover letter and resume, but also in any job interview you may secure. If you list a broad-based objective ("to have a satisfying career in cable television"), a cable company is going to think you're self-centered, caring more about pleasing yourself and being happy on the job than in helping them make money. If you list a very specific objective ("I want a job as a cable engineer"), they are likely to think that's exactly what you want. Maybe they don't have such a job, but they do have an opening for an installer—a line technician—a customer service representative. If their job doesn't match your objective, you may never hear from them. In fact, you may want to leave your job objective out of your resume.

PRESENTING YOURSELF AT YOUR BEST

Busy cable executives say they have no time to waste on applicants who haven't done their homework. If you're lucky enough to get an interview, *know* that cable station, the company, and the system.

As one cable executive puts it, "If you're not going to take time to investigate my company before you come into an interview, don't bother showing up. Whether you have potential or whether you don't, you've blown the interview."

Key personnel managers suggest you be prepared to talk about yourself and display your important skills. It's not bragging, they say. You need to demonstrate your accomplishments and to show that you have confidence in yourself. Use your enthusiasm and drive as you tell about things you've done in the past. But be sure you tie those accomplishments to a company's bottom line—not just, "I worked in sales for 6 months," but "Department profits rose xx percent." Emphasize your strengths. "Because I've demonstrated I can troubleshoot problems effectively, I believe I can give instructions to cable customers over the phone if they're having minor technical difficulties."

Do your best in an interview, even though there may not be jobs available at the time.

And network, Network, NETWORK! Keep a file of persons you've talked to. Stay in touch. Know if they change cable companies—where they've gone, and what they're doing. When you finish an interview, but you don't get a job offer, ask the manager if he or she knows anyone in cable whom you can call.

Don't give up. Personnel managers recommend you follow up on an interview (a thank-you note is always courteous) and continue to stay in touch with the company, phoning every three to four weeks.

IF YOU'RE A CABLE VETERAN

The job-hunting advice above still holds, even if you've previously worked in cable. However, if you're looking for your *second* (or third, fourth, or fifth) job in cable, you may be working with a search firm.

One international search firm that matches cable organizations' needs with cable veterans is Baker Scott & Co., Executive Search, a Parsippany, New Jersey-based company that specializes in cable recruitment. "We handle placement in accounting, finance, sales, ad sales, marketing, administration, operations, and general management," says Judy Bouer, a principal in the firm. "We basically do middle management-and-up for cable systems."

Bouer says her search firm needs to know the culture of a cable company or system before being able to come up with suitable candidates. "It's more than presenting the company with people who have skill levels," she explains. "We screen applicants first; then, for interviews, we send only those we think are right."

Cable companies or systems with jobs to fill contact Baker Scott & Co., she says. "They set parameters. They may want to see two, four, or six candidates for a single opening. What they want, we give them. And they pay the fees; applicants don't."

Eight out of ten cable jobs are never advertised, she warns.

> Yes, a company may post an opening on its bulletin boards, but there's a hidden, word-of-mouth job market.
>
> Even if a job is advertised, when someone is already working, they may not dare to answer an ad in the newspapers or trades. You don't know who'll get their hands on your resume. You can't take a chance.
>
> But when an applicant works with a company like ours, we don't let that happen. We know your name can't go all over the industry. Instead, we do a lot of reference checking—discreetly, so your own cable company doesn't find out you're interested in a new job.

Some job applicants contact Bouer's firm on their own, without waiting for a call from Bouer's recruiters. "If we can help, we do," she says. "Maybe we know a cable company in their area that's looking for someone. We can say, 'A super individual is available right in Philadelphia,' so the company won't have to pay the cost of relocation."

Bouer says any applicant she or her staff phone-interviews should have the basics about the cable system they're working for "down cold."

> For instance, you'd better be able to tell me how many subs you have. And if you're on the pay side, "In how many homes do you have your service? What is the cost?" If you're in operations, "How many people are on your staff? What do they each do?"
>
> We also ask situational questions. We need to know more than just skills—more than having someone say, "Yes, I do know how to put cable engineering together." We look for management and teamwork skills, too.
>
> I might tell an applicant, "Your boss gives you an assignment, and he says he wants it done by Friday. You give Betty the assignment. Friday, she tells you, 'Sorry, I didn't get it finished.'
>
> I ask the applicant, "What do you do? Who's taking the blame? Shouldn't you have checked up?" I want to see the applicant's reaction.
>
> Another situational question we use: "Your staff is standing on the other side of a wall and doesn't know you are there. What are they saying about you?"

Bouer and her firm check references thoroughly before passing applicants on to cable companies for interviews. "Some people are great actors," she warns. "We've caught people who have not really done the job they said they've done. Their resume looks wonderful on paper, but isn't true."

Because Baker Scott & Co. has been doing cable placement since 1980, she says, they have a network of contacts they can call to verify that candidates have been truthful. "Even if a cable company has gone out of business," she says, "we've been at this so long, that we generally know people who worked there. We can find them easily."

WHAT TO EXPECT

Corporate salaries are higher than field salaries, she says. "Out in the field, you're making less, but a corporate vice-president could be making six figures."

When her search firm places someone in a corporate cable position, Bouer says, it almost always negotiates a contract for the applicant that lists general (rather than day-to-day) responsibilities. The contract spells out answers to: What is your job? What are you expected to do? Are there stock options, perks, or other compensation arrangements, like bonuses?

Bouer prefers to work with applicants who have a minimum of three years experience in cable, but has placed persons who have not previously worked in the industry. "If a cable system needs a controller, and doesn't want to pay the expenses for relocating someone, we can often suggest someone who may fit."

While college is important, she says, cable systems are often willing to take persons who have 10–12 years of "terrific" experience. "For cable engineering, you need some kind of technical training," she says. "Speech degrees often help in getting marketing jobs.

"College helps you to think—gives you that little edge. Now that cable has become more sophisticated, we're seeing persons with master's degrees or MBA's."

Bouer's advice: "Take all the industry courses you can. Join organizations. See what's going on. Network whenever you can."

CHAPTER 11
NEW TECHNOLOGY

The combination of cable and telecommunications services is an alluring one, and technology is rapidly developing to make it possible. What this "marriage" means for future job prospects is unclear at present, though most cable veterans are predicting a job explosion. But one thing is certain. For many subscribers, cable ten years from now may offer far more services than they ever imagined.

NTIA Telecom 2000: Charting the Course for a New Century, a government publication, predicts a highly enhanced telecommunications and information infrastructure—connected by several alternative telecommunications and information systems which blend voice, data, and video communications. "As technology creates new services and competition drives prices down," the report says, "the prospects for a national or even international electronic neighborhood become greater."

FIBER-OPTIC CABLE

As franchises come up for renewal, many communities are requiring cable companies to rebuild and upgrade their systems. Many cable companies that are rebuilding plant are using fiber-optic cable to carry programming from their headends to intermediate distribution points, to interconnect cable systems, and to prepare for possibly eventually transmitting high-speed data communications.

For instance, between 1989 and 1992, Continental Cablevision rebuilt its five-system complex around Dayton, Ohio, using fiber to link its headends and create a regional telecommunications web. The result: an increase in the number of channels from 30 to 77, increased reliability, and improved picture quality. Continental's switch to the fiber loop gave the 5-system complex the capability to insert advertising on all 77 channels—up from the previous maximum of five. In addition, tying the various system segments together helped Continental to create a virtually seamless interconnect.

Some people have suggested that fiber-optic cable will reach the home for both television programming and telephone service—perhaps over a single integrated facility. While Continental had no specific plans for this in 1992, the rebuild also created a platform for alternate access phone service.

Alternate access is a phrase you'll be hearing more and more as the traditional lines between cable television and telephone service blur.

Because cable's traditional revenue streams are drying up, and because regulatory and legislative climates are changing, nontraditional revenue opportunities sound appealing to cable companies. For instance, in 1992, Tele-Communications Inc. and Cox (both cable giants) took control of Teleport, one of the two major providers of alternate access. An industry trade publication, *Cable World,* quoted Royce Holland, president of Metropolitan Fiber Systems, as saying, "Cable companies are getting very, very serious about getting into the telecommunications industry. This [the Cox buy] is a real wake-up call from them to the Bell operating companies."

INTERCONNECTS

Interconnect technology lets companies that market services to cable systems tie those systems together. Advances in that technology are producing more revenue.

Until recently, the large number of cable systems and the enormous amount of paperwork required have kept cable from realizing its revenue potential for spot advertising. Now, new technology allows ads to be beamed up to a satellite and back down to clusters of cable systems simultaneously.

In 1992, Cable Networks Inc., a national spot cable representative, began distributing ads by satellite to more than 45 cable systems in New York, New Jersey, and Connecticut. All were chained together by the New York Interconnect.

And to reach viewers economically and easily, CNBC, a satellite-delivered network, hired NuStar, a satellite ad service, to beam the same CNBC advertising spot to some 500 cable systems that reached 20 million homes. NuStar's parent company, the Lenfest Group, has also created AdStar, a satellite-delivery system, for national spot advertising.

Information Services

After the breakup of AT&T in 1984, telephone companies ("telcos") were initially forbidden to provide information services in their own operating areas. The *NTIA Telecom 2000* report, however, recommended that the Bell operating companies and other local telephone exchange companies be permitted to produce, store, disseminate, and process information. The report suggested that the federal government create policies and regulations for a telecommunications and information infrastructure—leading, perhaps, to the availability of "video dial tone" and other innovative services.

TECHNOLOGIES MERGE

No one is sure just what these exciting technologies will mean to the job market in cable television. But persons interested in cable television opportunities will almost certainly improve their chances for jobs if they also know something about telecommunications technology and how it is changing.

Learning computer technology also can help you see where cable is going—and may be part of the key to getting jobs in the exciting future.

Corporate giants like IBM and Time Warner have begun to explore merging data transmission technology and media (including cable) technology in a synergistic venture. They plan to use IBM technology to store movie images in compressed form, and then use advanced IBM transmission technologies to send them to any home that requests them through Time Warner's cable network.

The system also lets cable viewers use their TV sets interactively. Theoretically, they can play video games against other people located anywhere in the U.S., or answer questions on issues raised during a show.

Other companies are exploring high-tech transmission systems and interactive television. For instance, in 1992, Hewlett-Packard Co. brought out a device placed on top of the TV set that let home viewers take part in interactive programs, as well as send data from pocket computers.

An industry trade publication, *Cable World,* predicts that "the migration of much of the telecommunications business to digital technology will eliminate barriers between industry sectors, and expand the prospects of synergies between the cable and computer industries."

ANALOG AND DIGITAL TECHNOLOGY

To understand the new technology, and what it means to future jobs in cable, you need to know a little bit about two different ways of sending signals.

Analog is a term used for a signal that's a continuous variable. You can understand what the term means if you think of the way in which plants, animals, or children grow. Little by little, day by day, they increase in size. The changes happen—but you can't identify them from one moment to the next. You don't say that a baby weighs 12 pounds at 4 P.M. on June 24th and 12 pounds 1/4 oz. fifteen minutes later.

Digital, however, is a term used for a signal encoded as a series of discrete numbers. You can understand the term if you think of counting eggs in an egg carton. You have exactly one, three, or seven eggs. You don't have one-and-a-little-bit-more eggs. Digital signals don't degrade because of variations in signal strength in the way that analog signals do. Rich Prodan, director of CableLabs' advanced television laboratory, predicts that digital video offers cable greater reliability and lower overall costs in the long run.

Telephone—and cable—services began as analog technology. Both are changing as rapidly as possible to digital technology. On the way: Integrated Services Digital Networks (ISDNs)—which can use single equipment to offer voice, data, and video services (including cable).

Soon to come, cable execs say, is digital compression—"dozens of channels of compressed digital video 'overlaid' on frequencies above a cable system's existing analog channels," as *Cable World* explains the technology. Eventually, digital will be adapted to carry High Definition Television (HDTV).

HIGH DEFINITION TELEVISION (HDTV)

High Definition Television (HDTV) is a generic term that describes a new generation of television receivers and production equipment. Compared to current television receivers, HDTV can provide better and clearer pictures, and a larger aspect (width-to-height) ratio.

In 1940, when television was just beginning, the National Television System Committee (NTSC) developed a technology for black-and-white television sets. NTSC technology was broadened later to include color and stereo sound. It uses 525 lines per frame, and has a 3:4 aspect ratio (height-to-width).

HDTV transmissions are expected to have over 1,125 lines per frame, (over twice as much information), and a 9:16 aspect ratio. Technologies have already been developed to compress the HDTV signal, digitize it, and expand it on the other end, at subscribers' sets, provided they own sets capable of receiving HDTV.

The *NTIA Telecom 2000* report says HDTV could revolutionize television. "In addition," the report continues, "if cable can provide this new technology significantly before broadcast television, there could be a fundamental change in the competitive environment of the video marketplace."

Clearly, then, you need to stay on top of what is happening. Read the trades, cable veterans advise. Find out all you can about new technical developments. Almost certainly, they will mean more openings for cable jobs. One of those jobs could be yours.

CHAPTER 12

WOMEN AND MINORITIES

Opportunities for women in cable may be greater than those in traditional broadcasting, many cable executives feel. "Any time you have a rapidly growing industry, it's a good opportunity for anyone to carve out a career," says Kay Koplovitz, president of USA Network, and perhaps cable television's top woman executive.

Her advice to women: "Don't be afraid to succeed at whatever level you want to achieve." She believes women sometimes feel more successful being second in line, under the command of somebody else. "Be willing to take responsibility as the final decision maker," she advises. "Don't be afraid of the top slot. Prepare yourself to do your job well. Meet your job responsibilities to the best of your abilities. Then look for ways of broadening your scope, taking on additional responsibilities, increasing your contacts."

Koplovitz suggests reading the trades, attending state and regional conferences, and networking. Women In Cable (See appendix A), a professional association, offers contacts and resources outside your company, she says. "We don't think of it as a job market. Instead, the organization does help make people more aware of their professional skills and contacts."

By 1989, cable's work force was 41 percent women, according to research by the Hudson Institute. Women accounted for 37.4 percent of cable firms' officials, managers, and other professionals; 7.3 percent of cable firms' technicians; 86.9 percent of their office and clerical staff; and 45.7 percent of their sales force.

MINORITY EMPLOYMENT

There's good news, too, for minorities who'd like careers in cable! Research by the Hudson Institute found that minorities represented 22.2 percent of the cable work force in 1989. "A breakdown shows that they accounted for 11.6 percent of the officials, managers, and other professionals in the cable industry; 19.2 percent of the technicians; 27.9 percent of the office and clerical staff; and 26.2 percent of the sales force."

Because of the changing demographics of the U.S. labor force, the cable industry is beginning to approach the issue of diversity head-on. "This can be accomplished," the Hudson Institute study says, "by implementing programs and holding workshops which address the differences that naturally exist between people of a variety of ethnic, racial, and socioeconomic backgrounds. And it means letting employees see how many creative business solutions can be developed when a team of very different people work together to tackle a challenge."

The changing demographics of many urban areas also mean that those who are fluent in a language other than English, as well having English skills, may find it easier to get into cable careers.

Data from the 1990 census show Hispanics make up nearly 10 percent of the nation's population. The 15 largest U.S. Hispanic markets (including Los Angeles, New York, Miami, Chicago, Houston, and San Antonio), represent more than 90 percent of the nation's Hispanic population. But individual markets differ.

In Los Angeles, Jesus Trevino, chairman of the Latino Committee of the Directors Guild of America, says Hispanics make up nearly 40 percent of the population. And that number continues to grow.

Census data (1990) show 4.8 million Hispanics in Los Angeles County and neighboring counties: Orange and Riverside—up 73.4 percent since 1980. New York has 2.8 million Hispanics, an increase of slightly more than 35 percent in a similar time frame.

Miami, with 1.1 million Hispanics, is up more than 70 percent. The Chicago metropolitan area, including Gary and Lake County, Ind., had 894,000 Hispanics, a 41.3 percent increase. The Dallas-Fort Worth area, with half a million Hispanics, is up 109 percent since 1980. And the state of Washington, with a quarter million Hispanics, records a whopping 136.7 percent increase since the 1980 Census.

Because cable television is market-driven, advertisers and producers are looking for strategies to reach this large audience of potential buyers. For instance, GalaVision, a Spanish-language cable service, often purchases programming that features Hispanic performance. The network buys feature-length films, sporting events, musicals, and documentaries.

The Miami-based Univision Spanish-language TV network, sold by Hallmark Cards Inc. in 1992 to a group including American, Mexican, and Venezuelan investors, reaches some 90 percent of the Hispanic TV households in the U.S. The network offers a variety of programming: from selected auto racing to talk shows like "Show de Cristina" and "Corte Tropical."

The Miami-based "Show de Cristina," broadcast in Spanish throughout the United States and in 15 Latin American countries, stars Cristina Saralegui. The show is also seen in English on CBS-owned TV stations in New York, Los Angeles, and Miami. A graduate of the University of Miami with a degree in mass communications, Saralegui spent 10 years as editor-in-chief of *Cosmopolitan En Espanol,* the Spanish-language version of *Cosmopolitan* magazine, which circulates throughout Latin America and the United States.

"Corte Tropical," the first Spanish-language sitcom on commercial TV, was created and written by Mimi Belt-Mendoza. Univision sends the weekly half-hour show to more than 400 affiliates in the U.S., and about a dozen countries in Latin America.

Hispanic market consumer and media patterns have become so important that in 1992, NuStats, Inc., a California-based marketing research firm, created Hispanic Infosource, the first single-source database to study the 25-million U.S. Hispanic market, which has a purchasing power of $173 billion. Charter subscribers included Univision, Telemundo, Pacific Bell, Hunt-Wesson, *Miami Herald/El Nuevo Herald,* and Tichenor Media Systems.

"Hispanic Infosource includes information useful for media planning and placement," says Dr. Carlos Arce, NuStats president. Among the data: print and broadcast media preference and usage.

INCREASING ROLE FOR WOMEN

Since the mid-1980s, the number of officials, managers, and other professionals employed by cable firms—approximately 18 percent of the cable work force—has increased by more than 30 percent, reflecting the growth pattern of the industry for those years. As women become a dominant force in the new worker category, their role in the day-to-day management and functioning of the cable industry will also grow, says the Hudson Institute.

The stories that follow of several women cable professionals reflect their diverse backgrounds, as well as their managerial skills and responsibilities.

A STATION GENERAL MANAGER

Over and over again, those in cable stress the need to be flexible—to work hard, to do the job that has to be done. That's certainly true of Susan Hook, general manager of WESTSTAR Cable Company, Inc., the cable station in Bishop, California.

Because Owens County, a four-hour drive from Reno and six-hour drive from Los Angeles, has just 18,000 residents in an area the size of Belgium, Hook's station has 120 miles of plant to serve her 6,000 customers. "Many cable systems have one headend," she says. "We have 12, because of our geography. We also have a fairly complex microwave broadcasting system, so we can receive Los Angeles signals."

Hook describes a recent day:

> I came in at 7 A.M.. The bills reached residents on Saturday, so we had two mailboxes filled up with payments needing processing.

My office manager was on jury duty, and my two front-office people were only partly trained.

I helped deal with the mail and answer questions from the front office. I had a prescheduled conference call with our corporate headquarters about the possibility of our taking over the production of commercials for other cable systems. I dealt with our accounting department about their questions on how to code invoices.

I sent in time sheets for payroll, and filled out a capitalized labor report—detailing improvements we'd made which added to our station's value.

I talked with a woman on our corporate office's marketing committee about ad sales on the corporate level, and how we'd evaluate different systems for incentive contests among our sales staff. I lobbied the corporate office for additional equipment.

We're looking at possibly changing our program guide, so I evaluated two or three guide services. I filled out a credit bureau's request to verify employment for one of our people. I talked to our chamber of commerce; I'm on a committee that's planning a trip.

My chief tech and I had lunch together. At the restaurant, we talked about what channels to use for our new packaged services when we increase our channel capacity, and our signal security. I spoke to the corporate office about their upcoming visit; sent them a list of employees, matched with social security numbers; and got a price on a truck from a local vendor. I talked to my production people about new equipment, and left the office at 5:30.

A 20-year cable veteran, Hook initially "signed on" when a cable station needed a clerical worker for office tasks while the regular employee took a lunch break. Despite working for a station, she didn't know what cable looked like for 18 months, she says, because at the time, she didn't live in an area serviced by cable. As a full-time office employee, she received and posted payments and filed paperwork. She learned to install cable. She began to work in marketing.

> As the system grew and was sold to a succession of owners, I worked my way into more interesting jobs. I became an administrator . . . I worked as the office manager. By 1986, the station

was bought by WESTSTAR—an MSO with customers in California, Idaho, Montana, and Nevada. I asked for a chance to be general manager. We developed our local programs. I worked out a projected business plan.

Hook points out that as the cable industry matures, service quality—including customer contact—becomes increasingly important in retaining customers. "It's expensive to wire the people who don't have cable yet," she says. "Instead, many cable operators are looking to increase revenue by selling more services: pay-per-view, and local advertising."

She feels women and minorities have excellent opportunities to succeed in cable. "The FCC has laws about equal opportunity hiring and monitors it closely," she says. "When a station has an opening, it's not enough for us to send out letters. If minorities don't respond, you've got to go looking for them." Her station actively seeks to hire native Americans—at 10 percent of the local population, the region's biggest minority.

"We look for a drug-free record," Hook says. "We test for alcohol and drug abuse. We look for a clean driving record. Our people have to be personable, bright, and willing to learn."

Hook's system normally promotes from within. When an engineer resigned recently, she promoted a technician to fill the spot, placing the tech on 90-day probation for his new responsibilities.

Entry-level jobs either in the office or in the field pay $6.50 an hour, Hook says. At the end of 90 days' probation, employees receive benefits. Medical, vision, dental, and life insurance are provided, along with a small raise. After that, employees are evaluated for raises annually unless they're promoted to a higher salary level.

Her advice: expect change—and be able to handle it.

> If you think you're going to get into a job where everything stays the same, forget cable. Cable is at the whim of politicians, both local and federal, as far as regulatory issues are concerned. Now, really exciting technology is opening up new opportunities!

Think of cable as a good corporate neighbor. You need to be sensitive to your community and to your customers. Seventy-five percent of what you sell is service—not television signals.

A REGIONAL DIRECTOR OF A PAY SERVICE

Regional director of The Disney Channel, Chicago-based Eva Dahm earned undergraduate and master's degrees in communications from the University of Iowa, but her career has been spent in marketing and advertising. Dahm did target marketing at the University of Iowa Press and worked in book publishing, then spent a year in hospital marketing at a time when her community was getting ready to launch a cable operation. "I was looking for a job," she recalls, "and they were looking for someone who knew about targeted marketing."

For seven years, she worked with American Television and Communications (owned by Time Warner); then, joined The Disney Channel. Her region includes the states of Illinois, Minnesota, Iowa, Nebraska, North Dakota, and South Dakota.

Here's how she describes her job:

> I have four marketing representatives who report to me as well as a secretary," she says. "We're responsible for making certain that The Disney Channel business in each of the cable systems increases in size.
>
> Our clients are general managers of roughly 1,100 cable television systems in those states. We encourage them to market our channel. We may tell them, "We've got a direct mail piece that we can customize and provide to you at no cost." They'll ask us whom we're targeting, and what their costs will be.

Dahm says until she got into cable, she didn't realize that people who hold positions as she does are "change agents" for the industry. "We can watch innovative programs to see what works—and what doesn't," she explains.

"We meet with marketing managers and general managers and say, 'Here is what you need to do with your business. Here's why.' "

Dahm sees her role as primarily one of strategic planning. "Although I'm talking marketing, I need to think strategically how to get a cable system to see one of our people. Do I call the general manager? How do we use the relationships we've developed? For instance, my representatives will talk to marketing managers, but I have to have a relationship with the system's general manager.

"How can we use The Disney Channel's resources to help the problem?" Courses in psychology help, Dahm says. "I've had a lot of my business experience in the sociology and psychology of organizations."

WOMEN IN CABLE

In addition to her responsibilities at The Disney Channel, Dahm is Region 5 director of Women in Cable (WIC), the national association. Women in Cable's mission is to empower women in cable and related industries to attain their personal, professional, and economic goals, and to influence the future shape of the industry. The association has four goals:

- To advance women in the cable industry by developing their leadership and management skills
- To be in the forefront of the industry as a respected advocate and catalyst for new productivity in a diverse and changing work force
- To encourage women to understand, celebrate, and communicate their career and life choices
- To establish a synergy between the chapters and national that ensures a strong and viable organization supporting WIC's mission.

In 1988, Women in Cable began examining trends affecting the cable industry. It identified the work force and its productivity as key elements impacting the industry's continued success.

Two years later, WIC commissioned the Hudson Institute to conduct a study of how current and future work force trends will affect the cable industry, and to suggest strategies for the industry to effectively prepare for the changes ahead. The study was released as a white paper and brochure: "CableForce 2000: the Work Force as a Strategic Resource."

The study indicated that future shifts in work force demographics will have a greater effect on the productivity of the cable industry than on many other industries. Specifically mentioned were the decline in the number of entry-level workers and an increase in the diversity of workers entering the labor pool.

Cable needs higher-than-average intellectual and reasoning abilities among entry-level service positions, the study said. The industry also needs higher technology, such as the installation of fiber-optic cables and computerized customer service systems.

More details on the findings are available in a 24-page brochure which can be purchased from Women in Cable Foundation, 500 North Michigan Avenue, Suite 1400, Chicago, IL 60611–3703. The complete study can also be purchased.

The study concluded that the cable industry's prospects for continued success will be determined by its ability to identify, attract, and retain high-quality employees who are able to provide current and future subscribers with an unmatched level of service.

Four key cable industry associations have joined forces to form the CableForce 2000 Alliance. Their objective is to let the cable industry know about the findings of Women In Cable's work force study. The alliance includes the National Association of Minorities in Cable (NAMIC), the Cable Television Public Affairs Association, the Cable Television Administration and Marketing Society, and Women in Cable.

A CABLE SYSTEM MANAGER

One member of the CableForce 2000 task force is Rhonda Christenson, general manager of Continental Cablevision of Northern Cook County. The system, in Chicago's northwest suburbs, carries 52 channels: two pay-per-view (PPV), seven premium services, and 43 network and satellite channels.

Like a number of those interviewed for this book, Christenson hadn't originally considered a career in cable.

I was good in math and sciences. I was majoring in physical therapy at the University of Wisconsin-La Crosse. But by my junior year, I was totally baffled by the many opportunities available. Every time I took a class in an unfamiliar subject, I was convinced *that* was the field for me.

Then I married and moved to Rockford, one of the larger cities in Illinois. From the day I started to work in cable, I knew what I wanted to do. I loved it!

Over an 11-year span, I worked my way up through the ranks. Eventually I became director of administration—a position with responsibilities for supervising customer service representatives (CSRs), dispatchers, all accounting functions, the management information system, and training. I ran the system during a transition year while it was changing hands.

By the time the second new owner came in, however, I'd been offered my present job, which is closer to Chicago. Our system has 47,000 customers in five Chicago northwest suburbs. We're part of Continental Cablevision, which has 2.8 million customers across the country, in more than 400 different communities.

Christenson says there's no *one* path for women who'd like a career in cable. On her staff, she has college graduates with degrees in business administration, marketing, television production, journalism, and government relations.

For her, the biggest thrill is the ongoing evolution of employee development—watching the process that trains employees to deliver excellent customer service. During her five-day week, she tries to stagger her schedules so she can interface with employees who work at different times. Some techs start at 7:30 A.M., she says; service techs are out in the field till 8 P.M.; and line techs are available 24 hours a day. Though she usually doesn't work on weekends, she tries to be on hand at headquarters for special projects, like big pay-per-view events.

"The challenging part of my job is to fully understand our market," she says. "I need to know what customers expect—then to find operational ways of meeting or exceeding their expectations. They want quality reception, signal reliability, and above all, responsiveness."

Christenson believes the rapid-paced change of cable television has made the industry an excellent field for women.

> First came technology because you had to build the cable system. Our next big phase was marketing because you needed to get customers. Next, came operations because you had to deliver good quality service and operate within budget constraints. Now, we're back in the technical stage.
>
> New "blue sky" opportunities are coming in fiber optics and digital compression. In a short period of time, cable has to acquire and retrain individuals who have the skills to jump in, to succeed in this fast-paced environment.

WALTER KAITZ FOUNDATION

As the Hudson Institute concluded, cable's record of minority hiring has lagged behind its efforts with women—possibly due to the industry's lack of visibility in the minority community and a lack of minority applicants. However, cable leaders are beginning to reach out to them in a proactive way.

In 1981, key leaders in the cable industry founded The Walter Kaitz Foundation, a nonprofit organization, as a living memorial to a cable television pioneer.

"Walter Kaitz was one of the true pioneers of the cable industry," says Foundation president Ruth Brumfield. "He founded the California cable trade industry association, which preceded the National Cable Television Association. As a cable lobbyist, he was instrumental in getting a lot of the California laws about cable enacted by the state legislature.

"Walter Kaitz acted as mentor to a number of people of color, bringing them into the cable industry."

Like Kaitz, the founders of the fellowship created in his honor recognized that the industry's ability to compete successfully hinged on building a strong, diverse work force reflective of the communities they serve.

The Foundation's goal is to recruit highly-qualified, experienced managers of diverse professional backgrounds into cable television.

"Since the first Fellowship class in 1983, the Foundation has brought over 100 talented people of color into the management ranks of the industry," Brumfield says.

Cable leaders who represent systems operators, programming services, and suppliers serve on the Board of Trustees, contribute funds, and provide expertise to the Foundation's program. Cable senior executives serve on the Operations Advisory committee, sponsor Cable Career Nights, screen applicants, take part as judges at selection days, serve as part-time faculty, and hire Walter Kaitz Fellows in the companies.

Over a thousand individuals from all over the country apply for a Kaitz Fellowship each year. Applicants must be members of an ethnic minority (as defined by the Federal Communications Commission), have a bachelor's degree, and a minimum of three years managerial or professional work experience. They must be willing to relocate.

About 10 percent of applicants make it through the comprehensive screening process and are invited to one of the four regional selection days. During each of these intensive two-day sessions, cable industry representatives evaluate the leadership potential of these experienced managers from a wide variety of disciplines.

Candidates are given simulated management exercises, Brumfield says. "They're ranked on quantified dimensions such as problem-solving skills, stress, oral communication, and sensitivity to diversity," she explains. "All judges are cable-industry representatives.

Cable and cable-related companies draw on this pool of talent throughout the year to fill management and professional positions. The Foundation works closely with the hiring companies to ensure the best match of talent and hiring needs.

All Kaitz Fellows are hired into existing full-time positions within the sponsoring companies.

A partial listing of opportunities available includes: management trainees, financial analysts, operation analysts, staff accountants, affiliate salespersons, controllers, marketing director, government and regulatory managers, account executives, human resource specialists, staff attorneys, and system engineers.

The Fellows are also sponsored by their companies to participate in the foundation's professional development program. This program supplements the company's on-the-job training, and ensures that the Fellows have the necessary cable background to excel in their new careers.

In four intensive three-day seminars, the Fellows are introduced to budgeting and financial analysis, management, marketing, customer service cable technology, and community relations. Within those seminars such topics as key ratios and finance concepts (industry-specific to cable), and career personal development, including "situational leadership" and "diversity training" are discussed in detail, Brumfield says.

For more information, write the Walter Kaitz Foundation, Preservation Park, 660 13th Street, Suite 200, Oakland, CA 94612. Recruitment and placement continue year-round.

A Kaitz Fellow

At the Chicago regional office of The Disney Channel, Dana Fujioka recently completed her year as a Kaitz Fellow. During that time, she worked as an affiliate marketing representative, as well as attending the Foundation's seminars.

Fujioka hadn't planned to enter cable at first. A graduate of the University of Colorado, she earned her bachelor's degree in business administration, with an emphasis on international business and marketing. Her first post-college job was with a Canon dealership in Denver. After a training program, she was promoted to management.

When the company downsized early in the 1990s, Fujioka wasn't sure what her next career move would be. "I did some consulting," she recalls. "I took courses in Lotus and WordPerfect to upgrade my skills. I'd talked to the people at Job Services, the Colorado state employment agency, about finding work, and they mentioned the Kaitz Foundation.

"Since I met the Foundation's four criteria, I submitted an essay. The screening continued, as they reduced the candidate lists to 17 or 18 people for a region."

Fujioka says she was delighted to make the career switch to cable. "I'd been considering cable as a possibility," she remembers. "In the cable industry, the sky is the limit, as far as opportunity goes.

"At that point in my life, I felt it was important to narrow my field . . . to begin to get industry expertise . . . to start building a reputation and value . . . to find an industry where my own personal skills could grow."

After she was chosen as a Kaitz Fellow, Fujioka was interviewed by several cable systems. "I wanted to hang my hat with the Disney Channel," she says. In her initial position as an affiliate marketing representative, she works with various cable operation systems in West Virginia and Michigan, spending about 60 percent of her time visiting systems there. "My job is to promote the Disney Channel and to facilitate subscriber growth," she explains.

"That means doing training at the individual systems . . . getting to know CSRs as well as management staff. Systems vary considerably. West Virginia systems have a tone, a character, a culture unlike that of Michigan or New York or California."

People skills, sales skills, and communications skills are vital for her job, she says. "You need to get the CSRs' attention. You have to be sure they know why it's important to market the Disney Channel."

Other skills she considers essential: time and territory management.

Fujioka feels opportunities for women and minorities exist in cable—more so than in some other industries.

> When I was looking for a job, before I was picked by the Kaitz Foundation, I asked myself, "Would my chances of rising through the ranks be encumbered?" I can't say it won't happen, but there seem to be fewer barriers for women and minorities in cable. Upward mobility in this industry depends a great deal on performance and competencies.

Her advice to young persons considering a cable career: be aware of what's happening in the industry, and do your job well. Reading the trades is crucial, she says. "You need to know who's who and what's what. You need to keep up with technology. You must talk to cable operators, because they shape the industry. You need to find out what direction they see the industry taking, and adapt to that direction."

Whether or not you can get a paid job in cable, she says, it's important for young people to find out what really goes on in the industry and whether they like it. She suggests working as a CSR, even during high school or college, if possible; securing an internship; or even doing volunteer work behind-the-scenes, perhaps helping to produce a program shown on a local cable system.

"Create opportunities for yourself, so you know what cable is all about," she recommends. "Then you'll have some sort of direction. Working in cable—whether or not you are paid for your first position—helps you to determine a career path, and to learn what you must do to be ready for your next step."

A NETWORK VICE-PRESIDENT

"I think the chances for minorities in cable television are good," says Curtis Symonds, vice-president, affiliate marketing and sales for Black Entertainment Television (BET). The first network to showcase quality black programming 24 hours a day, BET offers a unique selection of urban contemporary programming that includes music videos, sports, family sitcoms, concerts, specials, talk shows, children's programs, news and information. In 1991, BET's initial public offering of stock made it the first black-owned company to be traded on the New York Stock Exchange.

From Symonds' perspective, minority candidates need to take a proactive position in the hiring process. "You've got to take the initiative—to learn how to get into networking," he says. "For instance, if you're attending job fairs when you're graduating from college, you need to plan in advance for the arena. Get business cards from everyone you talk to, and try to jump on those opportunities."

Like many other cable executives, Symonds wasn't necessarily planning a career in cable when he graduated from high school. In fact, he was a physical education major at St. Joseph's College in Rensselaer, Indiana. Transferring to Wilberforce University, he ultimately graduated from Central State University, Wilberforce, Ohio, with a degree in physical education.

Four years later, Symonds joined cable as a regional marketing assistant. He worked as a regional manager with Continental Cablevision of Ohio, and eventually ran the cable system in Xenia, Ohio.

Symonds, who wanted "to get on the programming side," joined Group W cable in Chicago. Later, when Turner bought out the system, he went to ESPN and stayed for six years. "I started out in local advertising and became director of affiliate marketing," he recalls. From ESPN, he moved to BET network in 1988.

> When I came on board, our network was in 18.7 million homes. Four years later, our affiliates were 2,500 cable systems in the United States, Puerto Rico, and the U.S. Virgin Islands. We were in 33.4 million homes, according to Nielsen. My job is to get us into the 27 million homes where we don't yet have subscribers. We can do that by diversifying our programming . . . reaching out more.

Symonds says his change of heart on career choice isn't unusual in the cable industry.

"That's the beauty of cable," Symonds says. "I don't think you'll find too many people in this business who started out by studying cable. Instead, this business is relationship-driven. How you talk to people . . . how you handle people . . . that's the way you get accepted in cable."

Advice

His advice to others who'd like to break into cable: get involved early. Black Entertainment Television offers students opportunities to work in a cable environment, he says, "either in marketing or in production.

> We've hired a number of Howard University students who have come to us, working in-house in our internship program. Getting a start in cable early in your work life lets you get familiar with the people and procedures, and how things work. If you're famil-

iar with a person, you are more likely to get the job, when there's an opening.

Symonds warns that cable—like many industries—expects young persons, regardless of gender or ethnic background, to pay their dues by hard work.

> Sometimes people think they can come right out of college and make a million dollars, but that's a misconception. You've got to pay some dues to climb up the ladder. Cable people certainly don't all have college degrees—in fact, I have a friend who started out as a secretary and now runs a small MSO. She earned a degree after she began working in cable.
>
> If you probably went round to one out of any ten employees at Black Entertainment Television and asked them their backgrounds, I'll bet they may now have college degrees, but they started out in the mailroom. They've climbed to various jobs within our network. If you can live through the grunt work, showing your employer you really want the job . . . if you're willing to accept the challenge of doing your best wherever you are, knowing that if you hang in there, something good will happen to you down the road, you'll make it in cable.

Skills Symonds considers essential include time management and flexibility.

> In my day-to-day operations, I have to block off time to answer my calls. If you try to answer every call that comes through, when it comes through, you'll never get your work done. Young people need to know they'll be evaluated on their use of time. When they first get into cable, they say, "I know how to do this . . . it's easy." Then, several months later, they tell me, "It's unbelievable—the amount of work we have to do."

Flexibility, too, is crucial if you want to succeed in cable, Symonds explains. "You probably won't get something right away that relates to your college major," he says. "But whatever job you do land, do it well. Show your boss you're willing to work hard.

"Don't forget people skills, though." Symonds says little things count—like returning phone calls within a short period of time, or answering correspondence promptly.

> "Your response level is critical. Often, young people don't understand phone etiquette. I don't like to let a message sit around even for a day or two without my getting back to the person who sent it. I could have missed an opportunity, or ignored timely information.
> The bottom line is, you must set the climate for better relationships. My philosophy is to treat people the way you yourself want to be treated.

THE NATIONAL ASSOCIATION OF MINORITIES IN CABLE (NAMIC)

Symonds is a board member of The National Association of Minorities in Cable (NAMIC), an important trade association. Founded in 1980, NAMIC serves the unique concerns of minorities actively involved in the cable television industry. NAMIC focuses attention on the mutual interests of minority cable professionals and the cable industry at large.

NAMIC has sponsored business development symposiums to promote sound business relationships between minority entrepreneurs and cable operators. Since 1987, the organization has presented annual seminars on the challenges and opportunities of the urban market, in partnership with the National Cable Television Association.

The nonprofit association helps minority recruitment efforts for cable operators, programmers, and hardware manufacturers. It also sponsors management development seminars, focused on the diverse work force characteristic of cable's urban systems. NAMIC sponsors industry events with the Walter Kaitz Foundation and supports the Foundation's mission to expand minority employment throughout the industry.

Membership in NAMIC is open to individuals from all races and cultures, from every organizational level, and from a variety of businesses.

Two membership classifications are offered.

A *regular membership* is available to an individual who:

- is directly employed by a cable television operating company, program supplier or hardware supplier, or
- holds an ownership interest in a cable television operating company, program supplier, or hardware supplier, or
- is otherwise significantly and directly involved in the cable television industry.

An *associate membership* is available to the individual who is not directly involved in the cable television industry, but supports NAMIC's goals and programs.

NAMIC wants to accomplish the following goals: to encourage the full participation of minorities in the cable television industry; to foster the growth and development of its members and the cable television industry; and to provide service to its members as an industry trade organization.

All members belong to the national association and receive its benefits, including preferred rates for association-sponsored seminars and events, a subscription to *Spectrum,* the association's national newsletter, and opportunities to take part in a national network of minority business and cable professionals.

Chapters in Chicago; Colorado; Detroit; New York; Northern Florida; Southern California; Washington, D.C.; and at Pennsylvania State University, University Park, Pennsylvania provide opportunities for members to become involved with the community and to take part in professional development programs.

For more information, write National Association of Minorities in Cable, P.O. Box 3066, Cerritos, CA 90703–3066.

CHAPTER 13

INTERNATIONAL MARKETS

CABLE TELEVISION IN CANADA

The future looks bright for jobs in cable television, according to *Cablecaster,* Canada's cable magazine. Driving the job market will be technology, as Canada's cable systems consider carrying data transmissions.

A 1992 Canadian-government-sponsored study of future datacasting services showed that satellites and Vertical Blanking Interval (VBI) networks will be the preferred technologies for datacasting over wide geographic areas. The study also concluded that Canada's cable system will become the prime carrier for datacasting in Canada's urban markets.

VBI data broadcasting—datacasting, as it's known in the cable world—is a high-speed, one-way point-to-multipoint data transmission medium. In short, data can be transmitted very quickly via the television signal, sent from a single source point to many receiving locations. As Ted McClelland, marketing manager for Ottawa-based NORPAK Corporation, writes in *Cablecaster,* data can be carried over television networks with a standard television signal. The data can be received everywhere the signal reaches. Data broadcasting, McClelland says, offers cost-effectiveness, timely delivery, and comprehensive geographic coverage.

VBI data broadcasting (also called teletext) was developed in the early 1980s as a means of transmitting virtually any type of data along with the television signal. McClelland predicts the technology, well-established in Europe, will become widespread throughout North America. Here's how it works:

North American television signals have 525 horizontal lines, divided evenly into two fields. Each field has 262.5 lines. The first 21 lines of each field make up the Vertical Blanking Interval (VBI). You see this as a black stripe when your television set loses the ability to hold its vertical signal, and the picture starts rolling.

The black stripe is part of the video signal. However, it carries no information. It's black, because it's empty.

Within those 21 empty lines, lines 10 to 21 are available for transmitting data. Line 21 has been used for many years to transmit closed captioning. Viewers who are hearing-impaired receive the closed captioning through a special decoding box.

Since 89 percent of Canadian households are passed by cable, the VBI data broadcasting signals are available virtually everywhere in Canada—either by cable, or by satellite. Like a "letter" sent via electronic mail, each data transmission can be "addressed" to a single site, to a group or region, or to certain viewers.

Technology already exists that allows viewers to receive data transmissions in various ways. Data can be sent as database files or software, straight to a computer's memory. Graphics and data can be displayed on a video display terminal (computer monitor). Or documents can be sent straight to a viewer's printer.

If Canadian cable systems begin to use VBI data broadcasting technology extensively, they're likely to market the service primarily to government users and corporate users. For instance, financial institutions may want to send lists of "hot credit cards" directly to merchants.

Because many viewers can receive the data transmission simultaneously (just as many viewers watch a popular program on cable), VBI data broadcasting on the major national Canadian networks should become popular in the next 5–10 years.

Interactive Technology

Interactive technology has increased cable use and the enjoyment of cable services for Canadian subscribers who rent Videoway, a specialized unit offered by Montreal-based Videotron Itee. The unit is available to all Videotron's cable-television network subscribers, and is similar to the traditional remote-control channel selector. It consists of a cable convertor/decoder and the cordless remote control, with a keypad that lets users access all the Videoway services.

Subscribers with the unit can press various keys to use data banks, such as Home Shopping, Stock Exchange, Weather, Horoscope, Lottery, Shows and Restaurants. A second access key lets them play video games. Another key sequence lets them receive electronic messaging and mail.

A 1992 market research study of Videoway use conducted at the University of Montreal, combined with customer services data, showed that users averaged 11.5 hours per week of Videoway viewing time—including 5.5 hours on video games, 2 hours on videotex services. Eighty-seven percent of respondents said they would recommend Videoway.

Households that responded said 84 percent had subscribed to Videoway for video games; 29 percent for interactive events, primarily sports; and 27 percent for other videotex services besides games. More than seven out of ten subscribers consult the videotex services, and nearly 30 percent of them consult videotex daily. Most frequently consulted were the weather forecasts, lottery results, television listings, and the horoscope.

More than eight out of ten said they'd installed the terminal themselves, and found it easy to use the system.

Canadian Associations

To keep up with Canadian technical developments, and what they mean for cable jobs, you'll want to read the trades—especially *Cablecaster*. The Don Mills, Ontario-based magazine has been appointed official magazine of SCTE Canada, (the Society of Cable Television Engineers in Canada). It plans to cover SCTE seminars and meetings and publish technical papers.

Other associations you'll want to write for information include the Ontario Cable Television Programmers, the Ontario Cable Telecommunications Association, and the Canadian Cable Television Association (CCTA). Another nonprofit association, formed in 1991, is called Canadian Women in Radio and Television. Membership is open to both men and women.

CABLE TELEVISION IN EUROPE

Revenues from cable are rising impressively. London-based Kagan World Media predicted that 1992 consumer expenditures on cable and pay TV in Europe would reach $5.8 billion—up 25 percent over the preceding year.

Cable operators in Europe have been moving towards a high-definition TV (HDTV) standard, hoping to get it adopted and running by 1995. A major drawback is the expense of HDTV sets. One version requires a glass screen and tube. Less-expensive versions with flat-screen and liquid-crystal display were expected by 1994.

Even before HDTV's advent, Europeans tended to talk about "inferior" U.S. television pictures. That's because the U.S. standard is 525 lines of TV picture information, while the European standards (PAL or Secam) offer 625 lines.

U.S. technology for HDTV is being developed with digital standards, but the European version of HDTV (called HD-MAC) is an analog standard. HD-MAC offers 1,250 lines of TV picture information. Opportunities in television—both broadcast and cable—will depend, in part, on the standardizing of signals and the technical compatibility of programming.

European countries vary in their approach to cable television, especially in the extent to which government regulates the industry. In the Republic of Ireland, cable is price-regulated, and (in 1992) 40 percent of Cablelink Ltd., the largest cable operator, was held by RTE, the state

TV company. The other 60 percent of Cablelink was owned by Telecom Eireann, the country's telephone company.

Private broadcast television channels only became permitted in Spain in 1989. The TV ad market there was (in 1992) fragmented among two state channels that carry advertising, three private national channels, and eight regional channels.

Much of the success of cable television in Spain may hinge on whether the government will let state phone company Telefonica have a monopoly on the carrier service, or will let a second company provide it. Cable operators' association AESDICA estimated in 1992 that its members, who represent about 70 percent of Spain's cable sector, will invest $500 million annually between 1992 and 1996.

Unlike typical viewers in the United States, many Europeans regularly watch cable television programs that originate from countries other than their own.

In the Netherlands, for instance, 85 percent of residents subscribe to cable—but an ad-supported cable satellite channel from Luxembourg leads in Dutch ratings.

Eurosport, one of the two pan-European sports channels, says Germany and German-speaking countries are its biggest market, with nearly 14 million homes. The channel, launched by the French broadcaster TF1, is also seen in Britain and in Denmark.

Governmental Regulation in Europe

Opportunities in cable in Europe may be primarily in technical careers. However, the number of jobs—and their availability—may well hinge on *how* governments view cable and regulate it. Do governments think of cable, for instance, as a utility that should keep its prices artificially low? Or do they believe cable companies should generate a profit, as other commercial enterprises do?

In the Netherlands, for instance, Netherlands Cable Exploitation Co. (NKM), is gambling on new services to attract additional viewers. NKM, owned and funded by two big Dutch banks, has been testing digital audio service and a burglar alarm service, delivered through cable. An information programming and marketing service, shown on cable, lists all channels available in the Netherlands and carries cable advertising.

NKM plans to test pay-per-view movies offered by the Scandinavian film channel; an interactive teletext service that allows users of personal computers to interact with a 30,000-page databank through their cable TV connection; and a monitoring service through a utility company that will tie cable subscriber's thermostats to a central computer. The computer can be programmed to turn heat off during the day and turn it on automatically at night.

In France, projected changes in cable operators' relationship with France Telecom should help lower subscription costs, *Cable World* projects. The French government hopes that Plan Cable systems will reach the break-even point by the end of 1995. In 1992, slightly more than half of France's 830,000 cable subscribers were on networks that are part of Plan Cable, the government project set up in 1982 to build cable. France Telecom constructed the network.

In an unrelated development, one proposal, backed by France's minister of posts and telecommunications, wants to let the two cable TV film channels show 500 films a year, up from the 364 allowed in 1992.

Cable in Moscow

Several Moscow organizations became part of a joint venture with Turner Broadcasting System in 1992 to start a 24-hour-a-day channel in Moscow that's predicted to eventually carry programming from CNN, TBS SuperStation, and Russian sources. Although CNN had been shown in Moscow on a 24-hour basis since 1991, most Moscow residents couldn't get it. That's because virtually all Russian TV sets only have six channels. Unless they upgraded their sets to receive UHF (ultra high frequency) signals, they couldn't see the CNN broadcasts.

The CNN/Moscow joint venture is being shown on a channel most Russian sets can receive. In 1992, two hours a day of the CNN news service were being translated into Russian, with the rest of the programming being shown in English.

Cable in the U.K.

Unlike cable television companies in the United States, cable companies in the United Kingdom are permitted to offer residential telephone service as well as cable services. The trend towards packaging telephone and cable TV as a joint sale is growing. Rory Cole, finance director of Videotron Corp., a partner with Bell Canada in London's cable franchises, says telephony doubles the cable company's capability to generate operating cash flows.

By summer, 1992, British Telecom, the telephone giant, said it was losing more than £1 million a quarter ($1.8 million in U.S. funds) to cable operators. As the company fought to keep its telephony market share in the U.K., *Cable World,* a trade magazine that regularly covers the international scene, reported a 17.4 percent penetration rate for residential telephone lines by United Artists Cable in Bristol, a town 120 miles southwest of London. In 1992, Birmingham Cable Communications announced it had sold over 8,000 telephone lines to residences—a 16.6 percent penetration of all homes eligible to receive the service.

Continental Research, a London-based market research firm, predicted that by the year 2000, telephony revenues could make up 40 percent of the total estimated cable TV revenues of £1.2 billion . . . rising from 10 percent of cable TV revenues in 1993 to 35 percent in 1997.

Current cable technology in Britain is involved not only with telephony, but also with signal encryption. In 1992, most British viewers were watching television through satellite dish receivers. That year, Continental Research estimated there were nearly 2.4 million dishes— one for every nine British homes. However, its study showed that in U.K. homes with satellite dishes, 50 percent of respondents preferred cable TV over a dish. *Cable World* said in 1992, direct-to-home dishes outnumbered cable subscribers eight to one. If signals aren't encrypted, dish owners can watch them free.

APPENDIX A

ASSOCIATIONS

Perhaps the leading source for cable television information is the National Cable Television Association (NCTA), the major trade association representing the cable television industry.

NCTA members operate cable systems serving over 90 percent of the nation's approximately 55 million cable television subscribers. NCTA also represents the cable industry's leading program networks, equipment manufacturers, and service firms.

Publications offered for sale by NCTA include *A Cable Primer, Careers in Cable,* and *Producers' Sourcebook,* a book that provides national and regional cable network guidelines for program acquisition. Technical papers from the annual convention technical sessions can be purchased, as can a 70-page glossary of Cable Television Technical Terms, developed by NCTA and the Society of Cable Television Engineers.

Founded in 1951, NCTA's principal mission is to provide its members with a strong national presence. NCTA represents the cable industry on Capitol Hill, in the courts, at the Federal Communications Commission, and with the national media. Over the years, NCTA has been instrumental in formulating and implementing policies and perceptions that affect the growth of the cable television industry.

For further information, write Association Affairs Department, National Cable Television Association, 1724 Massachusetts Avenue, NW, Washington, D.C. 20036.

Other associations or organizations related to the cable television industry include:

Association of Independent TV
  Stations, (INTV)
  1200 18th Street, NW, Suite 502
  Washington, DC 20036
Broadcast Cable Financial
  Management Association
  (BCFMA)
  700 Lee St., Suite 1010
  Des Plaines, IL 60016
  (For those already working in cable financial management and related areas.)
Cable Alliance for Education,
  (Cable in the Classroom)
  1900 North Beauregard Street, Suite 108
  Alexandria, VA 22311
CTAM (Cable Television
  Administration & Marketing
  Society)
  635 Slaters Lane #250
  Alexandria, VA 22314
Cabletelevision Advertising
  Bureau (CAB)
  757 Third Avenue, 5th Floor
  New York, NY 10017
CTIC (Cable Television
  Information Center)
  1700 Shaker Church Road, NW
  Olympia, WA 98502
CTPAA (Cable Television Public
  Affairs Association)
  414 Main Street
  Laurel, MD 20707
Communications Equity
  Association
  101 E. Kennedy Boulevard, Suite 300
  Tampa, FL 33602
CATA (Community Antenna TV
  Association)
  P.O. Box 1005, 3950 Chain Bridge Road
  Fairfax, VA 22030
FCC (Federal Communications
  Commission)
  1919 M Street NW
  Washington, DC 20554
International TV Association
  (ITVA)
  6311 N. O'Connor Road, LB-51
  Irving, TX 75039
MPAA (Motion Picture
  Association of America)
  1600 Eye Street NW
  Washington, DC 20006
National Academy of Cable
  Programming
  1724 Massachusetts Avenue NW
  Washington, DC 20036
National Association of
  Broadcasters (NAB)
  1771 N Street NW
  Washington, DC 20036
National Association of College
  Broadcasters (NACB)
  Brown University
  761 George Street
  Providence, RI 02906

National Association of
Minorities in Cable (NAMIC)
P.O. Box 3066
Cerritos, CA 90703-3066

National Association of
Telecommunications Officers &
Advisors (NATOA)
1301 Pennsylvania Avenue NW
6th Floor
Washington, DC 20004

National Association of TV
Producers & Executives
10100 Santa Monica
Boulevard, Suite 300
Los Angeles, CA 90067

National Cable TV Center &
Museum
Sparks Building Level B
Pennsylvania State University
University Park, PA 16802

National Cable TV Cooperative
14809 W. 95th Street
Lenexa, KS 66215

National Federation of Local
Cable Programmers (NFLCP)
P.O. Box 27290
Washington, DC 20038

National Telecommunications &
Information Administration
(NTIA)
U.S. Department of Commerce
14th & Constitution Avenue NW
Washington, DC 20230

Radio & TV News Directors
Association (RTNDA)
1000 Connecticut Avenue NW
Suite 615
Washington, DC 20036

Satellite Broadcasting &
Communications Association
(SBCA)
225 Reinekers Lane
Alexandria, VA 22314

Society of Cable TV Engineers
669 Exton Commons
Exton, PA 19341

Videotex Industry Association
8403 Colesville Road, Suite 865
Silver Spring, MD 20910

Wireless Cable Association
200 L Street NW, Suite 702
Washington, DC 20036

Women in Cable
500 N. Michigan Avenue,
Suite 1400
Chicago, IL 60611

APPENDIX B

RECOMMENDED READING AND RESOURCES

PERIODICALS

There are a number of periodicals connected with the cable television industry. Other periodicals of the film and video industry devote a great deal of space to covering cable television and its developments, especially from the business side.

Most of these are listed in *Ulrich's International Periodicals Directory,* a reference book that's available at many public libraries. You'll find some of them listed under the heading of "Motion pictures," while others are listed under "Communications—radio and television."

Another good source of periodical listings is *Audiovisual Market Place,* a multimedia annual guide published by R. R. Bowker Company, New York and London.

If you're interested in the technology of film and video production—HOW things are done—check *Applied Science and Technology Index,* a publication available on many libraries' reference shelves. Frequent updates will give you a list of articles in various technical magazines. Publishers' addresses are listed in the *Index.*

Write to the publisher, enclosing the price of a sample copy of the magazine, and ask to buy the back issue with the article you're looking for. This is a good way to learn about different technical developments, including what's happening in high-definition television research and applications, in digital compression, and in fiber optics—three technologies that are certain to impact cable television and career opportunities.

If you're especially interested in the business aspect of cable television, check the *Wall Street Journal Index,* as well as *Business Periodicals Index,* a similar index organized like the *Readers' Guide to Periodical Literature* or the *Applied Science and Technology Index.* Articles indexed generally deal with management issues, stock ownership, and industry trends.

Don't forget to check your own cable guide, if you're a cable television subscriber. If cities and towns nearby use a different cable company, read those cable guides too. You'll learn much from studying program listings, and from stories the guides run about networks and production specials.

Several television networks publish their own magazines. You can get subscription information and prices by listening to commercials aired on The Discovery Channel, and from Arts and Entertainment (for *The A&E Program Guide.* Subscribers to The Disney Channel also receive a magazine as part of their subscription fee.

For a comprehensive overview of educational programming that pulls together material from more than 30 broadcast and television works, you'll want to subscribe to *Cable in the Classroom.* This monthly magazine organizes program listings by subject matter and category for hundreds of educational programs, and includes information about conferences and special events.

The following periodicals are recommended:

Advertising Age
220 E. 42nd Street
New York, NY 10017

AdWeek
435 N. Michigan Avenue
Chicago, IL 60611

Broadcast Engineering
PO Box 12901
Overland Park, KS 66212

Broadcasting Magazine
1705 DeSales Street NW
Washington, DC 20036

Cable Avails
1905 Sherman Street
Denver, CO 80203

Cable Network Profiles (annual)
  Cabletelevision Ad Bureau
  757 Third Avenue
  New York, NY 10017
Cable World
  1905 Sherman Street
  Denver, CO 80203
Cableview Publications
  725 River Road
  Edgewater, NC 07030
Cablevision Magazine
  825 7th Avenue
  New York, NY 10019
CED
  P.O. Box 3043
  South Eastern, PA 19398
Communications Daily
  2115 Ward Court NW
  Washington, DC 20037
Communications Week
  1222 National Press Building
  Washington, DC 20045
CT Publications
  50 S. Steel Street, Suite 500
  Denver, CO 80204
Electronic Media
  740 N. Rush Street
  Chicago, IL 60611
Film & Video
  8455 Beverly Boulevard
  Suite 508
  Los Angeles, CA 90048
The Hollywood Reporter
  6715 Sunset Boulevard
  Hollywood, CA 90028
Information Gatekeepers
  214 Harvard Street
  Boston, MA 02134

Paul Kagan Associates Inc.
  (various newsletters, reports)
  126 Clock Tower Place
  Carmel, CA 93923
Lightwave
  P.O. Box 1260
  Tulsa, OK 74101
Media Business
  1786 Platte Street
  Denver, CO 80202
Multichannel News
  7 E. 12th Street
  New York, NY 10003
Nelson *Network*
  7 Crafts Road
  Gloucester, MA 01930
The 1992 Cable TV Fact Book
  (annual)
  Cabletelevision Ad Bureau
  757 Third Avenue
  New York, NY 10017
Private Cable
  1909 Avenue G
  Rosenberg, TX 77471
Producers' Sourcebook: A Guide
to Cable TV Program Buyers
  National Academy of Cable
  Programming
  1724 Massachusetts Avenue
  NW
  Washington, DC 20036
Satellite Business News
  1050 17th Street NW
  Washington, DC 20036
Satellite Communications
  6300 S. Syracuse Street,
  Suite 650
  Englewood, CO 80111
Satellite Orbit
  P.O. Box 10794
  Des Moines, IA 50340

SMMPTE Journal
595 W. Hartsdale Road
White Plains, NY 10607

Television Digest, Inc.
2915 Ward Court NW
Washington, DC 20037

TVSM
309 Lakeside Drive
Hursham, PA 19044

TV Technology
6827 Columbia Pike, Suite 310
Falls Church, VA 22041

Variety
475 Park Avenue S.
New York, NY 10016

Because cable television technology and telecommunications technology are becoming closer, the following periodicals covering telecommunications news are also recommended:

AT&T Technical Journal
AT&T Bell Labs
101 John F. Kennedy Parkway
Short Hills, NJ 07078

Business Communications Review
BCR Enterprises
950 York Road
Hinsdale, IL 60521

Communications News
124 S. 1st Street
Geneva, IL 60134

Fiberoptic Product News
Gordon Publications, Inc.
13 Emery Avenue
Randolph, NJ 07869-1380

ISDN Report
Probe Research, Inc.
Three Wing Drive, Suite 240
Cedar Knolls, NJ 07927-1097

Satellite Week
Warren Publishing
2115 Ward Court NW
Washington, DC 20037

Telecommunications
685 Canton Street
Norwood, MA 02062

Telephony
Telephony Division,
Intertec Publishing
55 E. Jackson Boulevard
Chicago, IL 60604-4188

BOOKS

The following books are recommended.

Cable TV Hardware & Technology, New York. Frost and Sullivan, 1987.

Cable Yellow Pages, Englewood, CO. Media Image Corporation, 1990. (annual)

The Hollywood Reporter Blu-book Directory. Hollywood, CA. The Hollywood Reporter. (annual)

Internships 1992: 50,000 On-the-Job Training Opportunities for College Students and Adults. Princeton, NJ. Peterson's Guides, 1991.

Television and Cable Factbook: 1990 Edition, Warren, MI. Warren Book Publishing, 1989.

Bartlett, Eugene R. *Cable Television Technology and Operations.* New York. McGraw-Hill, 1990.

Baylin, Frank et al. *Off-Air Satellite & SMATV: The Private Cable Multi-unit Handbook,* Boulder, CO. Baylin Publishing, 1987.

Baylin, Frank and Gale, Brent, *Satellite & Cable TV: Scrambling and Unscrambling,* 2nd ed. Boulder, CO. Baylin Publishing, 1988.

Bone, Jan. *Opportunities in Film Careers,* 2nd ed. Lincolnwood, IL. VGM Career Horizons, 1990.

Carter, T. Barton et al. *The First Amendment and the Fifth Estate: Regulation of Electronic Mass Media,* Westbury, NY. Foundation Press, 1989.

Deschler, K. *Television Technology.* New York. McGraw-Hill, 1987.

Dominick, Joe. *Broadcasting Cable & Beyond.* New York. McGraw-Hill, 1990.

Draigh, David. *Behind the Screen: The American Museum of the Moving Image Guide to Who Does What in Motion Pictures and Television.* New York. Abbeville Press, 1988.

Eastman, Susan T. et al. *Broadcast-Cable Programming: Strategies and Practices.* 3rd ed. Belmont, CA. Wadsworth Publishing, 1989.

Fry, Ronald W. *Internships, Vol. 1: Advertising, Marketing, Public Relations, & Sales,* 2nd ed. Hawthorne, NJ. Career Press, 1990.

Fry, Ronald W. *Internships, Vol. 5: Radio & Television Broadcasting & Production,* Hawthorne, NJ. Career Press, Inc., 1991.

Jones, Glenn R., *Jones Dictionary of Cable Television Terminology, Including Related Computer & Satellite Definitions,* Englewood, CO. Jones Twenty-First Century Ltd., 1988.

Jurek, Ken, *Careers in Video: Getting Ahead in Professional Television.* White Plains, NY. Knowledge Industry Publications, 1989.

Kensinger, Jones, et al. *Cable Advertising: New Ways to New Business.* NY. Prentice-Hall, 1986.

Oringel, Robert S. & Buske, Susan M., *Access Manager's Handbook: A Guide for Managing Community Television*. Stoneham, MA. Focal Printing, 1987.

Piti, Theresa, editor. *Backstage TV, Film and Tape Production Directory*. NY: Backstage Publications Inc. (annual)

Sautter, Carl. *How to Sell Your Screenplay: The Real Rules of Film and Television*. NY: New Chapter Press, 1988.

Taylor, Tom T. III et al, *AgeWise: A Case Study of a Public Access Cable Television Program*. Portland, OR. AgeWise Publishing, 1989.

Warner, Charles H. & Buchman, Joseph. *Broadcast & Cable Selling*, 2nd ed., Belmont, CA. Wadsworth Publishing, 1991.

Warren Publishing, Inc. staff, *The Business Television Directory, 1991*. Warren, MI. Warren Book Publishing, 1990.

Wiese, Michael, *The Independent Film and Videomakers' Guide*, Studio City, CA. Michael Wiese Productions, rev. ed., 1990.

APPENDIX C

GLOSSARY

Below is a list of terms commonly used in cable television.

Access channels: Channels which a cable system has designated for local use. Some channels may be allocated to the government, some to educational institutions, and some to other local organizations. Public access channels are reserved for the use of individuals and groups within the community.

Addressability: The ability of a cable company to direct programming to a subscriber's home on demand. If a cable subscriber has an addressable convertor, the company can easily make changes in service (or discontinue premium service) without having to physically come to the customer's home.

Aerial plant: Cable suspended in the air, strung between telephone or utility poles.

Alphanumeric: The capability of putting characters such as alphabet and numbers on the television screen.

Amplifier: Sometimes called a booster. It makes the electronic signal stronger as it travels long distances throughout the cable system.

Analog: Signals that are continuous variable sounds.

Arbitron: One of the rating services that measures viewing.

Area of Dominant Influence (ADI): Arbitron's term for how far the broadcast signal reaches in a given market area.

Aspect ratio: The ratio of height-to-width in television. In the United States, broadcast television sets have traditionally had a 3:4 aspect ratio. Sets that can receive High Definition Television signals are expected to have a 9:16 aspect ratio.

Avails: The amount of time in which advertising can be sold on cable channels.

Basic service: The type of service or selection of services that a local cable operator provides at the least expensive price. Generally, if you're a basic cable subscriber, you are receiving a combination of channels and program offerings that your cable company has packaged together for one monthly cost. Basic service does not include the so-called "premium" channels, for which you pay extra.

Cable: The coaxial cable that carries the television signals. They can come from broadcast television stations or from satellite signals.

Cable information services: Services that could be delivered through telephone or cable television networks, such as bill-paying or other banking services, and consumer videotex.

CATV: Community Antenna Television, or cable TV.

Channel capacity: How many channels a cable system can carry at one time. Earlier systems had much less capacity than those being installed today.

Churn: The turnover as subscribers disconnect and reconnect.

Coaxial cable: A common means of transporting a signal in cable television. It's a central core, surrounded by an outer conductor. Because of coaxial cable's broadband capability, it can transmit a large number of television signals at the same time.

Convertor: A small box that is generally placed on top of the television set is supplied by the cable company. It converts the cable television signal into separate, or "discrete," channels that the home TV can receive. Older convertors frequently had a device called a pron, with identification for the specific customer coded in. If the subscriber wanted to add or subtract extra services, the cable company sent a serviceperson to change the pron. Newer convertors are addressable; that is, service changes can be made directly from the cable company's office without a home visit.

Customer Service Representative (CSR): A person who works for a cable system and answers questions from customers, often by telephone. Questions can be on billing, on the quality of cable service, or on problems such as outages. CSRs often take orders for additional cable channels, and may actively market cable services.

Daypart: The time segments that comprise a typical day of TV viewing. For example, 7 P.M. to 8 P.M. is "prime access" and 11 P.M. to 11:30 P.M. is "late news."

Demographics: An analysis of consumers or audience, based on economic and social background.

Designated Market Area (DMA): Nielsen's term for how far a broadcast signal can reach in a given marketplace.

Digital: A signal encoded as a series of discrete numbers. Many cable and telephone systems are upgrading to digital technology—virtually a "must" in today's computer-driven world.

Digital compression: The ability to compress television signals together so as to have the capabilities of increased channels with the same amount of "space." Digital compression is a new technology that will give cable systems the ability to upgrade. Systems can turn one video channel into additional channels (perhaps as much as four times the channel's original capacity) without having to physically rebuild the cable system.
Disconnect: Cancelling cable services.
Distant signals: TV signals that originate away from the subscriber's area. "SuperStations" of Atlanta (WTBS), Chicago (WGN), or New York (WOR) are classified as distant signals when they are viewed in another city.
Drop: The hookup between a cable system and the individual subscriber's TV.
Drop cable: The piece of cable that directly enters the home of a subscriber.
Earth station, earth receive station, downlink. An antenna shaped like a dish that receives signals from a satellite.
Enhanced Redundancy Protection: Backup protection, so cable signals won't go off the air if there's a technical problem.
Feed: The source of programming or the television signal.
Feeder line: The piece of cable that connects the main trunk line or cable to the smaller drop cable.
Fiber optics: A different kind of cable than coaxial. This alternative technology transmits cable TV signals and uses glass-like threads that are thin as hair.
Franchise: An agreement by which a municipality grants a cable company the right to construct a system.
Hardware: The physical equipment for producing, storing, distributing, or receiving electronic signals. For cable television, hardware includes the headend, the coaxial (or fiber-optic) cable network, amplifiers, television receivers, and equipment needed for production: cameras, videotape recorders, etc.
Headend: The central receiving location of a cable system. This includes the community antenna, downlink, and equipment necessary for satellite reception, amplification, and transmission of the signal to subscribers and other transmission points within the community.
HBO: Home Box Office. The first pay-TV movie network.
High Definition Television (HDTV): A television technology whose characteristics are still being developed, but which is expected to produce a picture with greatly improved quality. US HDTV transmissions are expected to have 1,125 lines per frame (over twice as much information as signals using technology developed by the National Television System Committee) and a 9:16 aspect ratio.

Homes passed: The total number of homes which could be hooked up to a cable system, regardless of whether they actually are.

Independent: A cable system that is not part of an MSO (multiple system operator).

Information Retrieval: Any of several systems that allow users to obtain information via one-way or two-way cable. (See teletext and videotex.)

Information Services: Services cable subscribers can potentially buy and receive through their television sets: access to banking, computer information, etc.

Insertion: In cable advertising, picking and choosing programming in order to place a commercial in a given time break.

Interactive cable: Also called two-way cable. Cable systems and hardware that allow two-way communications. Downstream communications go from the headend to the subscriber's television set. Upstream communications carry information from the home to the system headend. Upstream communications are only possible in two-way systems.

Interactive video training: Training delivered by television, videotape, or laser disc in which the answers and questions change, depending on the response of the viewer.

Interconnect: A way in which two or more cable systems link themselves together to distribute a commercial advertising schedule simultaneously. Interconnects help make an advertising schedule more effective. Advertisers only need to negotiate a single contract with the interconnect, rather than having to buy time themselves on more than one cable system.

Leased channel: A channel for which the cable operator charges a fee.

Local access: The "right" of a community to have locally-produced programs shown on cable. Frequently negotiated as part of a franchise.

Local origination: Similar to local access, but may also describe local programming which the cable company itself produces, specifically for a particular system or municipality.

MATV: Master Antenna Television System. A system which uses one central antenna for picking up broadcast signals, and is frequently used in an apartment high-rise.

MDS: Multipoint Distribution System. Commonly used to bring pay TV to hotels, this service uses a very high frequency to transmit one television signal.

Microwave: A broadcasted line-of-sight transmission at frequencies well above the normal TV frequencies.

MSO: Multiple System Operator. A company that operates more than one cable television system.

Multiplexing: The ability of a single cable channel to receive different feeds, so it can present different programs at different times. From a cable operator's

view, multiplexing adds value and choice because stations can present more product during prime viewing.
Narrowcast, narrowcasting: The direction of programming to a selected, high-interest audience.
New build/rebuild: A cable system under construction.
Nielsen: A rating service.
Pay programming: Special programs for which a charge is made in addition to the basic cable fee. These often include movies, sports, or made-for-cable specials.
Pay-per-view: A service made possible by the development of addressable technology that lets the subscriber order specific shows on a pay-per-event basis. These could include major motion pictures, sporting events, or specials.
RFP: Request for Proposal: The document issued by a municipality to those who apply for a cable franchise. It sets forth the municipality's standards of selection.
Satellite: A device which is in stationary orbit. A communications satellite (such as SATCOM, COMSTAR, WESTAR) serves as a repeater. It receives signals from uplinks on earth and relays them back to cable systems through downlinks on the ground.
Software: Programming and materials, such as films, videotapes, and slides.
Spots: Individual advertisements appearing on cable; commercials.
Subscription television (STV). Pay-TV programs for which signals are scrambled. A special receiver at the subscriber's set decodes the signals.
Subscriber: A person who pays for cable services.
SuperStation: Television stations that are syndicated via satellite to cable systems throughout the country.
Teletext: A one-way information retrieval system that lets users obtain information. Users choose from a variety of printed material, including headlines, news, financial data, schedules, and listings.
Tiers: Additional levels of cable service which the subscriber can buy for a charge above that of basic service.
Transponder: The device on a communications satellite which permits signals to be transmitted and received.
Trunk line: The major cable used to distribute the signal. Trunk lines divide into feeder lines. Feeder lines are connected to drop cables, which go directly to a subscriber's home.
Underground installation: Burying of cable. Cable can also be installed on poles (aerial suspension).
Vertical blanking interval: In North American television technology, the first 21 horizontal lines of each field of a TV signal. Lines 10 to 21 can be used for transmitting data as part of the television signal. Like a "letter" sent via

electronic mail, each data transmission sent by the VBI can be "addressed" to a single site, to a group or region, or to certain viewers.

Videotape: A method of recording sight and sound electronically. Videotape can be erased and re-recorded. It can be instantly played back.

Videotape Recorder (VTR): The device which records the videotape.

Videotex: Two-way information retrieval, possible with interactive service. Videotex gives access on demand to additional information bases, as well as services like home banking and home shopping.

VGM CAREER BOOKS

OPPORTUNITIES IN
Available in both paperback and hardbound editions
Accounting
Acting
Advertising
Aerospace
Agriculture
Airline
Animal and Pet Care
Architecture
Automotive Service
Banking
Beauty Culture
Biological Sciences
Biotechnology
Book Publishing
Broadcasting
Building Construction Trades
Business Communication
Business Management
Cable Television
Carpentry
Chemical Engineering
Chemistry
Child Care
Chiropractic Health Care
Civil Engineering
Cleaning Service
Commercial Art and Graphic Design
Computer Aided Design and Computer Aided Mfg.
Computer Maintenance
Computer Science
Counseling & Development
Crafts
Culinary
Customer Service
Dance
Data Processing
Dental Care
Direct Marketing
Drafting
Electrical Trades
Electronic and Electrical Engineering
Electronics
Energy
Engineering
Engineering Technology
Environmental
Eye Care
Fashion
Fast Food
Federal Government
Film
Financial
Fire Protection Services
Fitness
Food Services
Foreign Language
Forestry
Gerontology
Government Service
Graphic Communications
Health and Medical
High Tech
Home Economics
Hospital Administration
Hotel & Motel Management
Human Resources Management Careers
Information Systems
Insurance
Interior Design
International Business
Journalism
Laser Technology
Law
Law Enforcement and Criminal Justice
Library and Information Science
Machine Trades
Magazine Publishing
Management
Marine & Maritime
Marketing
Materials Science
Mechanical Engineering
Medical Technology
Metalworking
Microelectronics
Military
Modeling
Music
Newspaper Publishing
Nursing
Nutrition
Occupational Therapy
Office Occupations
Opticianry
Optometry
Packaging Science
Paralegal Careers
Paramedical Careers
Part-time & Summer Jobs
Performing Arts
Petroleum
Pharmacy
Photography
Physical Therapy
Physician
Plastics
Plumbing & Pipe Fitting
Podiatric Medicine
Postal Service
Printing
Property Management
Psychiatry
Psychology
Public Health
Public Relations
Purchasing
Real Estate
Recreation and Leisure
Refrigeration and Air Conditioning
Religious Service
Restaurant
Retailing
Robotics
Sales
Sales & Marketing
Secretarial
Securities
Social Science
Social Work
Speech-Language Pathology
Sports & Athletics
Sports Medicine
State and Local Government
Teaching
Technical Communications
Telecommunications
Television and Video
Theatrical Design & Production
Transportation
Travel
Trucking
Veterinary Medicine
Visual Arts
Vocational and Technical
Warehousing
Waste Management
Welding
Word Processing
Writing
Your Own Service Business

CAREERS IN Accounting; Advertising; Business; Communications; Computers; Education; Engineering; Health Care; High Tech; Law; Marketing; Medicine; Science

CAREER DIRECTORIES
Careers Encyclopedia
Dictionary of Occupational Titles
Occupational Outlook Handbook

CAREER PLANNING
Admissions Guide to Selective Business Schools
Career Planning and Development for College Students and Recent Graduates
Careers Checklists
Careers for Animal Lovers
Careers for Bookworms
Careers for Culture Lovers
Careers for Foreign Language Aficionados
Careers for Good Samaritans
Careers for Gourmets
Careers for Nature Lovers
Careers for Numbers Crunchers
Careers for Sports Nuts
Careers for Travel Buffs
Guide to Basic Resume Writing
Handbook of Business and Management Careers
Handbook of Health Care Careers
Handbook of Scientific and Technical Careers
How to Change Your Career
How to Choose the Right Career
How to Get and Keep Your First Job
How to Get into the Right Law School
How to Get People to Do Things Your Way
How to Have a Winning Job Interview
How to Land a Better Job
How to Make the Right Career Moves
How to Market Your College Degree
How to Prepare a *Curriculum Vitae*
How to Prepare for College
How to Run Your Own Home Business
How to Succeed in Collge
How to Succeed in High School
How to Write a Winning Resume
Joyce Lain Kennedy's Career Book
Planning Your Career of Tomorrow
Planning Your College Education
Planning Your Military Career
Planning Your Young Child's Education
Resumes for Advertising Careers
Resumes for College Students & Recent Graduates
Resumes for Communications Careers
Resumes for Education Careers
Resumes for High School Graduates
Resumes for High Tech Careers
Resumes for Sales and Marketing Careers
Successful Interviewing for College Seniors

SURVIVAL GUIDES
Dropping Out or Hanging In
High School Survival Guide
College Survival Guide

VGM Career Horizons
a division of *NTC Publishing Group*
4255 West Touhy Avenue
Lincolnwood, Illinois 60646-1975